# Colour
# for Survival

# Colour
# for Survival

Peter Ward

Orbis Publishing · London

Printed in Italy

ISBN 0 85613 093 1

Frontispiece: Herd of roan antelopes near Lake Chad. Among large ungulates the need for camouflage is greatest at dawn and dusk when big cats are on the prowl

Right: Stinging caterpillar showing its warning colours

# Contents

# Colour in the Animal World

Colour has a crucial role to play in the defence strategies of many animals, apart from being important in their social lives and in their relations with plants. Most animals have enemies against which they must defend themselves if they are to survive and reproduce. Only large creatures such as whales and elephants, and predators at the top of the food-chain like sharks, lions and eagles are free from natural enemies – and even they need protection when they are small.

Basically, there are four ways an animal can gain protection: it can conceal itself; it can flee, if it is fast enough and can sense the danger in time; it can stand its ground and rely on whatever weapons it has; or it can bluff the predator into thinking it has a powerful defence when actually it has none. Some animals have specialized greatly in one of these strategies, but most keep their options open to some extent. They select the best strategy for a particular situation, but if it fails they have another on which to fall back.

Species without a weapon of any kind, and some that do have one but are 'reluctant' to use it, often rely on concealment to escape their enemies. Small animals can simply hide in a crevice or other inaccessible place. This may not be as safe as it seems, however, for if reached by a predator there is little chance of escape. Instead, many stay in the open, concealed by

A tropical American bush-cricket
*Tanusia* resting among mouldy
leaves which it resembles.
'Disguise' is concealment that
depends on similarity to familiar
inanimate objects. It is a system
that demands stillness and a
highly-modified body-shape.

camouflage or disguise and prepared to adopt a second defensive tactic if approached too closely.

The term 'camouflage' (alternatively 'crypsis') is properly restricted to cases where the animal simply blends with its background, and does not resemble any other object. 'Disguise' refers to situations where the animal looks like some inanimate object such as a leaf, stick, stone or bird-dropping. To a predator, a camouflaged animal is difficult to see because its outline is obscure. A disguised animal, on the other hand, is easily seen but looks like something inedible.

Two examples will make this distinction clear: a female pheasant, crouched on her nest in a tangle of vegetation, goes unnoticed until almost walked upon. This is camouflage, for it depends entirely on the variegated pattern of the plumage blending with the background. An Indian leaf-insect is also difficult to locate, but in this case the animal actually resembles the leaves on which it feeds. Its green, leathery wings have a leaf-like form and bear a perfect representation of the midrib and veins, and its thorax and legs are flattened to simulate smaller leaves. This is disguise, for the isolated insect looks like a bunch of leaves.

These are selected examples and it is not always easy to decide whether an animal fits best into the camouflage or disguise category. Presumably most disguised animals passed through a stage when they were merely camouflaged in the course of their evolution. Some present-day species are still at an intermediate stage.

Generally speaking, animals protected by concealment are relatively defenceless species which many predators find palatable. But not all animals are easy to kill and good to eat. Quite a large number have extremely un-

pleasant attributes such as distasteful, or even poisonous, chemicals in their bodies, irritant hairs or spines. Some have a potent weapon such as a sting or poison fangs, and others are protected by impregnable armour. Well-defended creatures such as these generally advertise the fact. Otherwise they risk being damaged before predators have had time to discover the unpleasant aspect of their intended victim. Advertisement is usually achieved by a conspicuous show of bright colours, combined with some form of extrovert behaviour. Red, yellow and black are well-known 'warning colours', as evidenced by hornets, wasps and bees, but metallic blues and greens, and even white, may serve the same purpose. For it is not the colours themselves that are important but the degree to which they render the animal conspicuous. Not all conspicuously coloured animals, however, are unpalatable; bees and wasps, for instance, have been copied by harmless hover-flies, which, like many other mimics, have evolved not only the same

Above: Powerful animals without natural enemies, like the black rhino, have not evolved protective coloration. The baby stays close to its mother at all times.

Left: 'Camouflage', shown by a baby East African ostrich, requires an appropriate colour pattern, but no modification of body-form.

colourful patterns as their models but also their shape and behaviour.

Unfortunately the term mimicry has been used rather loosely in the past. Stick- and leaf-insects, for example, are often referred to as mimics, which confuses the sense of the word in a discussion on animal coloration. The copying of inanimate objects is best referred to as 'disguise', while 'mimicry' is restricted to cases where one species imitates another. This is not just a semantic distinction, for it is obviously convenient to have one word applicable where predators are tricked into not noticing their prey and another for situations where they notice it but are tricked into leaving it alone.

Animals using colour for defence mostly fall into one of the above categories, but as one might expect there are numerous exceptions. For instance, some species are well camouflaged or disguised while at rest but flash previously-hidden warning colours the moment they are disturbed. Whether the warning is real (asso-

Right: 'Warning Colours' blatantly exhibited by a poisonous arrow-poison frog, *Dendrobates pumilio*, from Costa Rica.

Below: In simple 'Mimicry' harmless animals copy the warning colours of well-defended species. The European hover-fly *Syrphus batteatus* superficially resembles a wasp, but has no sting.

Above: The secretary-bird is a specialist snake-hunter. Such keen-sighted diurnal predators have obliged many animals to evolve elaborate protective coloration.

ciated with a weapon), or a bluff intimidation display, there can be no doubt that the defence strategy has suddenly changed. Such multiple defence systems are commonplace, especially in the humid tropics.

Defence strategy may also alter with age. Young animals may rely greatly on concealment when they are small and helpless but assume warning colours as they develop a more potent defence later in life. Conversely, some warningly-coloured youngsters grow up into cryptic adults.

Colour, like beauty, is in the eye of the beholder. All animals are coloured, but the colour has become important in defence only where it can be seen. Permanently subterranean species, living in caves or beneath the soil, and those living in the abyssal depths of the oceans, cannot be expected to display adaptive coloration – except possibly where they produce their own 'chemical light'.

Since colour can be appreciated only by predators with well-developed vision, it can be involved only in defence against vertebrate predators (fishes, amphibians, reptiles, birds and mammals), or the few invertebrates, such as squids, with acute vision. Colour obviously plays no part in defence against enemies that hunt by smell, touch or sound. The African striped mouse is well camouflaged among dry grass stems, and thereby gains protection from falcons, jackals and cats which hunt by day or moonlight. But its camouflage affords little protection from marauding barn owls, which can hunt by sound alone, nor from pit-vipers guided by the radiant heat of the mouse's body.

Predators also may need to be inconspicuous, especially if they hunt by stealth. Large carnivores such as the python, tiger and pike, have clearly evolved camouflage to improve their efficiency as killers. By allowing undetected approach it helps shorten the final crucial dash at the victim. With smaller predators the situation is sometimes ambiguous. The beautiful disguise of a flower mantis, and the adjustable camouflage of the chameleon, may well be of assistance when they are hunting, but their crypsis is probably far more important in protecting them from their own enemies, particularly birds.

We are considering here the ways in which colour is used in conflicts between species, but colour often has an equally important part to play in the relationships between members of the same species. This is the social role of colour. Bright colours feature in the sexual displays of many animals, and are often combined with elaborate ornamentation that heightens the effect. Communication between a young animal and its parents may also involve colour. Many birds are stimulated to place food inside the beaks of their nestlings by the conspicuous pattern inside their mouth openings. Another social use of colour is in alerting the members of a group to danger. An alarmed rabbit or deer raises its tail to reveal a prominent white patch as it dives for cover. Other individuals react instinctively to this signal without

waiting to see for themselves the source of the alarm.

Bright colour used for social ends can obviously pose problems, for it makes the wearer conspicuous to its enemies also. This risk has been overcome, or at least minimized, in a number of ways. Animals that rely greatly on crypsis generally keep any colourful patches well-hidden when they are not required. As every ornithologist knows, many of the dull-coloured birds that are difficult to identify when at rest or feeding quietly can be distinguished easily in flight. White patches above the tail, conspicuous wing-bars and other markings then become glaringly obvious. And because these patterns are used for courtship and other displays between members of the same species they frequently differ greatly between species that look very similar when resting. Similarly, butterflies often have their brilliant colours confined to the upperside of the wings, and are cryptic when the wings are folded. Bright colours show also in flight, of course, but then the insects' vulnerability to predators is far less. Alternatively the bright colours may be worn obviously, but only during the breeding season,

Below: Bright colours required for social display can easily spoil an effective camouflage. The Malaysian flying dragon *Draco volans* blends with the tree-bark on which it hunts. By inflating its brilliant yellow throat sac it can nonetheless perform a spectacular social display.

to minimize the risk. Sticklebacks and newts, for example, show their nuptial colours for only a few weeks in spring, while many sandpipers, African weaver-finches and other birds moult into a distinctive breeding plumage which they retain for a few months only. It is only the male who compromises his camouflage by wearing bright colours in this way. The female, who plays a more crucial role in reproduction, remains cryptic throughout. It is interesting to note that in the odd cases where sexual roles have been partially reversed it is the male who is permanently camouflaged. For instance, the female painted snipe of tropical Africa is brightly coloured and actively courts the drab male. She abandons the male after laying, leaving him to incubate the eggs and rear the brood alone.

Before going on to look in detail at the ways animals use colour in their relations with other species it is useful to consider what colour is. Sunlight is formed from a mixture of different colours, as we see when a rainbow forms during a shower. The raindrops act as tiny prisms, splitting the light into its component colours, each of a different wavelength. Surfaces that reflect all wavelengths in sunlight appear white, while those that absorb look black. Most objects absorb certain wavelengths, and reflect the remainder. Since we see only the reflected light, this determines the object's colour. For instance, a vermilion flycatcher looks red because its plumage absorbs most of the purple, blue, green, yellow and orange wavelengths, and reflects only red light towards our eyes.

There are two ways in which animal colours are produced: one involves pigments, chemicals that have the capacity to absorb certain wavelengths and reflect others; the other is by the physical splitting of white light into its spectral components which then travel in different directions. The splitting results from physical phenomena that occur when white light shines on finely textured surfaces or through thin transparent layers. Colour produced this way is termed 'structural'.

Biochemists have now been able to extract hundreds of chemically different pigments from animals, and many more remain to be described. Perhaps the most ubiquitous biochromes (as the pigments of animals and plants are called) are the carotenoids. These are mainly responsible for red, orange and yellow colours. Apart from their use as pigments, carotenoids have an important physiological role connected with vitamin A formation. This is one reason for the abundance of carotenoids inside many animals where their colour cannot normally be seen. Familiar examples are the pink flesh of the salmon and the yellow yolks of birds' eggs. In view of their importance it is curious that these pigments cannot be synthesized by any animal,

though plants make them in great abundance. Herbivores obtain them directly from their food while carnivores must either consume herbivores or eat some plant material themselves. Domestic cats may often be seen eating grass, and the facile explanation that 'they do it to make themselves sick' is widely accepted. More likely they are extracting plant juices before regurgitating the fibre.

Another widespread group of pigments are the melanins, which appear black, brown or occasionally dark-red. Unlike carotenoids, melanins can be synthesized by animals. They are generally made in the skin and tend to be concentrated in a layer beneath the surface, or in structures such as hair, feathers or scales. The all-white animals (and humans) with pink eyes, called 'albinos', are the result of an occasional genetic mutation. They are completely unable to produce melanin and, unless cared for specially, are doomed. Their eyes are devoid of pigment and derive their colour from the blood vessels which show through the transparent tissues. In consequence their sight is extremely poor.

The majority of white-coated animals, however, are not albinos. Their eyes are normally pigmented and they often have some patches of dark colour. Such 'leukemic' mutants are generally selected out of the population as fast as they appear, for they are extremely conspicuous in most situations. However, in areas where snow forms the usual background leukemism has been favoured and there are white populations of many birds and mammals. The opposite condition, 'melanism', refers to an excessive production of the black pigment. Melanistic animals also originate as occasional mutants and are generally at a disadvantage, but in some situations they, too, have been favoured. Black 'panthers' arise in all leopard populations, but in open habitats their hunting ability is impaired by their colour, so they die out. In the dark jungles of south-east Asia, on the other hand, the black individuals do best.

Other classes of pigments – quinones, pterins, flavins and many more – are for the most part responsible for white, yellow and red coloration. Blue pigments are rare, though blue is common as a structural colour. Bilin is the pigment responsible for the blue-green colour of many birds' eggs, while green is usually produced by combination of a yellow pigment with a structurally produced blue.

Structural colours are often produced by interference, a physical effect which occurs when light passes through thin, transparent layers. Interference colours are to be seen on a soap bubble and in the shimmering hues of a dragonfly's wings. As long as there are spaces between successive layers, even quite thick structures can produce interference colours.

Left: Iridescent colours are produced when light passes through a thin transparent membrane. The bright colours of these glasswing butterflies mating in a Central American jungle advertise their noxious attributes.

The iridescent colours of many beetles and butterflies are created this way. In the beetle's wing-cases, and in the minute scales that clothe the butterfly's wings, are numerous thin layers of transparent material, separated by films of air. Exceptionally, the layers are separated not by air but by a liquid. For this reason, specimens of tortoise beetles (family Cassidae) lose their fabulous pure-gold look as they dry, to the chagrin of many a collector. The metallic colours of hummingbirds, crows and other birds are also produced by interference. The thin layers responsible are in the barbs of the feathers which are made of sheets of transparent keratin separated by air; the pigment layer is beneath.

Structural colours are produced also by Tyndall scattering, named after British physicist John Tyndall. When white light passes through a fluid, or transparent solid, in which extremely fine particles are suspended, some of the light hits the particles and is scattered. The blue wavelengths are scattered more than the red-and-yellow components of the light so the whole medium takes on a blue colour. John Tyndall discovered the phenomenon when seeking to explain why the sky is blue. He found that fine particles of dust in the atmosphere scatter the different wavelengths of light that comprise sunlight to varying degrees. The shorter wavelengths at the blue end of the spectrum are scattered most and so reach our eyes as we look

Left: Structural colours are also produced when sunlight is reflected from a multilayer structure containing films of air. In this way many beetles acquire a jewel-like appearance which warns birds of their poisonous nature.

Right: The five-spot burnet belongs to a small group of European moths that can fly by day, protected from birds by toxic chemicals in their bodies. The metallic green is produced by physical effects; the red spots are patches of non-iridescent scales containing a red pigment.

upwards. Many birds, reptiles and fishes are coloured by 'Tyndall blue', in this case microscopic particles embedded in the transparent keratin layers being responsible. The blueness is often enhanced by the scattered red and yellow wavelengths being absorbed by an underlying layer of pigment.

Whiteness, strictly speaking the total reflection of all colours, is also produced by physical effects. Although composed of ice, which is a transparent substance, snow looks white because air trapped between the crystals, and the numerous reflecting surfaces, bounce the light back unchanged. In a similar way white feathers derive their whiteness from air bubbles embedded in layers of transparent keratin. If the reflecting surfaces are arranged regularly they act like a mirror, producing a silvery lustre. Crystals of guanine in fishes' scales have such a regular structure.

As we have seen, animal colours mostly derive from sunlight, but in the sea there are many creatures that make their own light. This 'bioluminescence', as it is properly called, is produced by complex chemical reactions that occur in the luminous organs. The result is a continuous or pulsed emission of an eerie light, usually green or yellow in colour. Bioluminescence has been aquired mostly by deep-sea animals, though there are some surface-living examples. The planktonic organism, *Noctiluca*, is responsible for the beautiful glittering surf on tropical beaches. There are few luminescent species in terrestrial habitats, though fireflies

and glow-worms are plentiful in warm regions. There are also cave-dwelling insects that produce light. Luminescent organs are used mainly for social signalling, but they may have a further role in defence. The detailed behaviour of deep-sea animals, many of them festooned with coloured lights, is still poorly understood, though one of the squids is known to shoot a luminescent cloud into the face of an attacker in the dark. This is the equivalent of the sepia ink 'smoke-screen' of squids in surface waters.

Most animals have their external coloration fixed, at least for quite long periods. They can change it only when shedding the skin or, in the case of mammals and birds, by moulting — the procedure for renewing fur or feathers which may take weeks or months to accomplish. A variety of animals can alter their coloration, without skin changes, over a period of days, but a few only, such as *Anolis* lizards and fishes like the plaice, can perform the rapid colour-changes associated with chameleons. While frogs and toads undergo colour-changes also, they are generally much slower. All these lower vertebrates share a common technique for colour adjustment. Arranged as a superficial layer in the skin are special cells, called 'melanophores', loaded with fine granules of the black pigment melanin. Within each melanophore the granules can spread and retract under nervous control. When the melanin granules are fully dispersed they absorb all the light, and the skin appears dark; when they are pulled in, lighter-coloured pigments beneath prevail.

Right: The West African
chameleon adjusts the shade of
its green body to that of the
surroundings by subtle changes
in a black pigment layer in the
skin. When the black
'melanophores' are contracted
the underlying green colour
predominates. The melanophores
expand completely when, caught
in the open, the chameleon
feigns death (far right).

Squids and octopuses are the most spectacular quick-change artists, able to range over a variety of colours in a matter of seconds. In the skin of these advanced molluscs are 'chromatophore' cells containing different pigments. These colour-bearing cells can, in various permutations, expand and contract rapidly, each controlled by a tiny muscle. Most species have chromatophores of three colours, often red, yellow and black, but a layer of reflecting cells, 'iridocytes', modifies the effect of the pigments and increases the range of colours.

The slow colour-changes of spiders and caterpillars are unexciting in comparison. They may take days to accomplish, and merely require changes in the proportions of different skin pigments.

Colour-change is of considerable value to cryptic species since it allows them to match their appearance to that of the background. But it may also be involved in social displays. The common African agama lizard, *Agama agama*, is generally brown, like the ground or tree bark on which it lives. Only the dominant male in each group stands out because of his vivid blue-and-orange nuptial dress. The bright colours serve to attract the females in his harem, to show his dominance over young males, and influence disputes with the dominant males in adjacent territories. If he is cornered by a predator, however, he changes rapidly to the drab, cryptic colours worn by his fellows.

Should he succumb in the attack another male in the group soon assumes the conspicuous coloration that goes with the top-male position.

Colour-change may also be used as threat display. The common African chameleon, *Chameleo dilepsis*, employs subtle colour adjustment constantly as it moves between different backgrounds, but when molested it performs a dramatic open-mouthed threat-display accompanied by much hissing. As it does so, its pattern changes and black stripes and spots appear that are not shown at other times. In this case the display is mainly bluff, but the flashing colour-change of certain venomous octopuses signifies a very real threat, and inquisitive predators do well to take heed.

Many reptiles and amphibians also use their ability to adjust skin colour for thermo-regulation. Indeed this could well be the original function of the melanophores, colour-change for defensive and social purposes being later elaborations. Lizards and toads, including the European common toad, *Bufo bufo*, often bask in the early-morning sun, warming their bodies and thereby increasing their metabolic rate. Often they become much darker when they are sunning. This is because the melanophore cells in the skin are expanded, allowing the granules of pigment to intercept as many as possible of the sun's rays. Melanin absorbs infra-red wavelengths (radiant heat) just as efficiently as it does the wavelengths of visible light.

# Camouflage

By definition a well-camouflaged animal is difficult to find largely because its overall coloration matches that of the background. But there is usually more to it than that. For instance, it may carry a confusing pattern that renders the body-shape difficult to make out. Often, too, the animal is unusually flat, or gives the illusion of flatness even though its body is rounded quite normally. More remarkably, such giveaway features as eyes and legs, and even the animal's shadow, can be made inconspicuous. These are some of the anatomical and colour adaptations connected with camouflage, but behavioural modifications are demanded also. Animals that rely greatly on camouflage for defence tend to feed, and carry out other activities, at night. Throughout the day they remain perfectly still in a carefully selected resting place. The many thousands of moth species are mainly in this category, for most are palatable and much sought after by birds. The few day-flying moths are distasteful (and warningly-coloured), like the burnets and foresters (Zygaenidae), or mimics of wasps such as the clearwings (Aegeriidae). There are also a few particularly agile flyers like the humming-bird hawk-moth, which are able to evade their enemies while visiting flowers during the day. Most of the other camouflaged animals that move about by day, usually because only then can they find food, have a tendency to

The Texas horned lizard shows many of the characteristics associated with camouflage: a squat form reducing contour and shadow; an irregular outline; fine markings that blend with the background; and a 'disruptive' dorsal-stripe 'cutting' the body into halves.

walk slowly and warily. And they are always prepared to 'freeze' or take to flight at the first sign of danger.

The requirements for good camouflage can be met in many different ways, even by members of the same group. Conversely, unrelated animals have often evolved similar concealing patterns that suit a particular background. For example, a grasshopper living among grass-stems may have the same striped pattern as the spiders and mantises living in the same vegetation, and look very different from a close relative inhabiting a pile of rocks nearby. Such 'convergent evolution' can make it difficult to grasp the fundamental importance of various adaptation by looking at animals group by group. A much clearer picture emerges if we concentrate instead on the ways in which the five basic requirements for effective camouflage are met, drawing examples from as wide a variety of families and habitats as possible. The five requirements for camouflage are: background colour-matching; disruptive patterns; flattening, optical or real; concealment of eyes, limbs and shadows; and appropriate behaviour. Some animals display all five, others are able to dispense with one or more because of special circumstances. Species which rest in dimly-lit places, for example, have no problem with shadows.

The simplest kind of background-matching, in fact the simplest camouflage of all, is shown by animals which rest on uniformly coloured backgrounds such as sand, mud or snow. Even the surface of a single leaf may provide a plain background for diminutive creatures such as aphids and smaller tree-frogs.

The one plain background that might seem to defy natural selection is clear water. Yet in both freshwater and the sea there are many

animals that match the watery surroundings. The glass catfish, *Kryptopterus bicirrhis*, an open-water species of tropical Asian rivers, has achieved the desired result through transparency, as its name implies. Its thin, delicate skin, muscles, and even skeleton, are all translucent. There are also transparent creatures in the sea, including eggs and larvae of many fishes as well as a great variety of jelly-like invertebrates. The larger, pelagic (open-sea) fishes with strong skeletons, powerful muscles and stout integuments cannot be transparent. Some, like tuna and mackerel, are blue or blue-green above, like the water beneath them, and silver below. The silver belly renders them invisible to sharks, barracuda and other predators looking up toward the mirror-like surface of the ocean. The small fishes living in huge shoals near the surface (herrings, shad, sprats) are silvery all over. This protects them from predatory fishes, but may leave them open to attack from above. Terns, gulls and other sea-birds gather in large numbers to prey on the small fish should they venture too near the surface. Fishes that live in the surface layers of lakes and rivers are mostly plain-coloured also. This applies to warm-water as well as cold-water species, though one gains a somewhat different impression of the fishes of the Amazon, Congo and other large rivers when looking at tropical-fish tanks. Fish-fanciers have naturally favoured the strongly patterned and colourful species, many of which come from small tributaries, while rejecting the drab varieties.

Another simple coloration is uniform whiteness. White animals occur in all sorts of habitats and without some knowledge of their natural history one can easily obtain wrong notions concerning the value of their simple coloration. In fact white is probably the most difficult coloration of all to interpret. Brown and green animals are generally cryptic, while red, yellow, black and metallic colours usually signify advertisement — for social or defensive reasons; but white can mean anything. Surprisingly, even in the forest it can be cryptic. For example, there are white mantids and crab spiders that sit on flowers of the same colour awaiting their prey and obviously well-camouflaged. Alternatively, white can be for warning. The cabbage white butterfly, *Pieris brassicae*, is known to be highly distasteful to birds, like others among 'the whites'. Finally, an all-white appearance may have nothing to do with defence, serving instead as a highly visible social signal. Gulls, terns, pelicans and egrets are probably white so as to be conspicuous to members of their own species, for they follow one another to food sources.

There is one habitat, however, where white animals predominate and where whiteness is clearly for camouflage. This is, of course, in the

Arctic, Antarctic and on snow-covered mountains. The smaller Arctic birds and mammals, like snow buntings, ptarmigan, lemmings and Arctic fox, are white only in the winter months. In spring they moult into sombre brown coats that provide concealment after the snow has melted. Some of the larger species remain white throughout the year, as will be discussed later.

The other major biome (major category of habitats) in which pale animals predominate is, of course, desert. Typical desert species are sandy or tawny, but it is not quite true to say that there are no colourful animals. For when it rains, as it does from time to time in even the most arid areas, the flush of ground vegetation

Left: A long-horned grasshopper changing its skin after dusk. Species reliant on camouflage must perform any conspicuous behaviour at night.

Right: White-tailed ptarmigan in the Rocky Mountains. Whiteness for camouflage is mainly confined to arctic and alpine species.

Below: Glass catfish, one of the few vertebrates using transparency for defence.

holds an abundance of green insects, including caterpillars and grasshoppers. Colourful birds also move in to take advantage of the copious insect supply. But for most of the year — and sometimes for years on end — the active forms of life in the arid landscape are mostly those coloured to match the ground on which they live.

In the Sahara Desert, characteristic mammals are the camel, gazelles, fennec, jerboa and gerbils — all basically sand-coloured. Resident birds such as the desert eagle owl, golden and Egyptian nightjars, sand-grouse and larks, and reptiles like the horned viper and desert monitor lizard share the same general tone. The exceptions are black millepedes and darkling ground beetles, which are warningly-coloured because they are distasteful. It might seem incongruous that desert crows and ravens are also black, but these belligerent birds have no need for camouflage, and require conspicuousness as an aid to flock cohesion.

In the deserts of western North America live an equally wide collection of desert-toned species including the pronghorn antelope, kit fox, desert cottontail rabbit, rattlesnakes and many lizards. The chain of deserts running from western Canada to central Mexico have formed in regions of quite different geology and illustrate well the principal of ecological variation in cryptic animals. Pocket-mice and deer-mice in the related genera *Perognathus* and *Peromyscus* are particularly instructive, for these small, ground-living rodents have

evolved to suit a wide range of backgrounds. In typical sandy areas they are fawn-coloured like most desert mammals, but in areas of volcanic rock in the Mojave Desert, where the ground is black, the same mice are darkly pigmented. Conversely, on the dunes formed from gypsum-derived sand in New Mexico, known as White Sands, lives an almost white mouse, *Perognathus gypsi*, which is not found anywhere else.

We may presume that in all these desert areas mice of different colours are arising constantly as a result of mutations and recom-

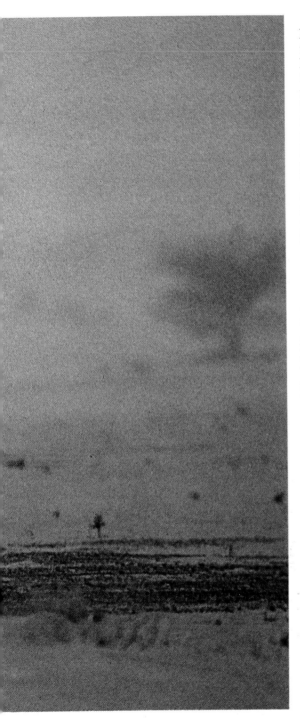

The house sparrow, *Passer domesticus*, is known as the English sparrow in North America, though it actually originated somewhere in the Middle East and has long been established throughout Europe as a ubiquitous commensal of man. In 1852 it was first introduced into the New World and from the original small colony in New York city it has spread far and wide. Now, little more than a century later, sparrows living in the arid south-west are significantly paler than those on the eastern seaboard. Whether natural selection has brought this differentiation about because paler sparrows are better camouflaged in the desert, or for physiological reasons as some believe, matters little to us here: either way it shows that the coloration of a population can become adapted to a local situation in a matter of centuries only.

For many of the creatures that feed or rest among fresh leaves an overall green coloration provides simple but effective camouflage. In tropical rainforests, where the green is perennial, there are many green frogs, snakes, lizards, birds and insects. But in the drier tropics and temperate regions green coloration is less widespread. It is confined mainly to those insects and spiders that are active only in summer, and to birds like the warblers that spend only the summer months in the deciduous forests and winter in the tropics.

Since leaves have prominent venation and the surfaces are often strongly textured, a plain-green coloration is suited only to a few frogs, insects and other small creatures. Exceptional among the larger animals are long snakes like the African green mamba, *Dendroaspis*, and the American 'green snakes', *Opheodrys*, which can make themselves look small by coiling. The majority of green-forest species show patches of other colours. Among the birds, Amazon parrots and parakeets are the most familiar, but there are also brilliant-green broadbills, motmots, jacamars, bee-eaters, and others. Most of them have some red, yellow or other contrasting colour on the body somewhere. At close quarters these patches are conspicuous and may be used socially, but from a distance, in the gloom of the forest, they are not. Indeed they help the camouflage by breaking the body outline.

In the sea, green coloration for concealment is restricted to shallow waters where there are seaweeds. The beautiful sea-slug, *Elysia viridis*, is common on rocky shores round Britain, yet so well does its bright green coloration match that of the sea-lettuce (a seaweed) on which it feeds that it generally goes unnoticed.

So far the backgrounds considered have been uniform, but for species which habitually rest among dead leaves, or on tree-bark, coarse-grained rock or other variegated surfaces, simple colour-matching is not sufficient. 'Back-

Left above: Dorcas gazelle in a Saharan dust-storm. Sandy hues predominate among desert animals.

Left below: Many desert invertebrates, like this 'white-lady' spider from Niger, blend well with the sandy terrain.

binations. On light-coloured ground the dark individuals are conspicuous and soon killed, while on dark backgrounds it is the light ones that tend to die young. What biologists could only guess at until recently was how long it took for such adaptations to become fixed in a population. Most thought it about ten thousand years or more. The fossil record could provide no clues, since colours are not preserved. Recently, however, some indication of the time-scale has been gained from what may be called a 'natural experiment' and it is much less than was deduced previously.

Above: Ecological variation: The population of lesser earless lizards in the gypsum dunes of White Sands are all pale. More typical populations of the species (above right) living elsewhere in the southern United States have a variegated pattern suited to stony ground.

ground picturing' describes the patterns that blend with these more complex backgrounds. Only a few quite commonplace examples can be considered here; others may be found wherever one cares to search, on land or in the sea.

Tree-trunks are good hunting grounds for the naturalist interested in animal coloration. Commonest among camouflaged animals resting on the bark in most areas are moths. Many of them are patterned to resemble their background precisely and some take their name from the tree on which they habitually rest — pine hawk, willow beauty and alder, for example. The wings of the pine hawk, *Hyloicus pinastri*, are grey with rounded markings imitative of the scaly bark of conifers. Those of the willow beauty, *Boarmia gemmaria*, bear a delicately traced pattern resembling the fissured bark of willow trees. In warmer regions lizards and frogs, as well as insects of all kinds, rest on tree-bark and often share quite similar markings. Where the air is humid and trees are covered with lichen, many of the insects and spiders that rest on them are finely patterned with blue-green and brown. The beauty of some of the moths is recognized in names such as merveille-du-jour and oak beauty, two European species.

One of the moths that rests on tree trunks has become a biological 'classic' as a striking example of rapid evolutionary change. It has

Left below: Broad-billed motmot. Predominantly green birds are commonplace in tropical rain-forests, but rare in other habitats.

Right below: Green amphibians like this African tree-frog are also confined to evergreen habitats.

also provided experimental evidence of the extent to which a cryptic pattern protects the wearer from predators. The peppered moth, *Biston betularia*, is pale with fine black markings, rather like the bark of silver birch trees on which it rests during the day. To nineteenth-century lepidopterists in England it was a commonplace insect, but the all-black form that appeared occasionally among adults reared from batches of caterpillars was something of a rarity. Appropriately, as it turned out, they named it 'variety *carbonaria*'. By the beginning of the present century, however, the situation had become greatly altered. Var. *carbonaria* had

become abundant in northern England: around Manchester, for instance, only 5 per cent were of the 'normal' form.

It was not until the early 1950s that the entomologist H. B. Kettlewell was able to explain convincingly how and why the change in population had occurred. He began by suggesting that on the soot-blackened birches, victims of the Industrial Revolution, *carbonaria* individuals were better camouflaged than the normal form, whereas on the previously clean birches they had been at a considerable dis-advantage. Selective predation by insectivorous birds was the root cause of the rapid rise in the proportion of *carbonaria* individuals. Experiments of many kinds were performed with laboratory-bred moths of the two forms, peppered and black. In the simplest of the experiments moths were placed on clean birches away from the city, and on blackened trees in an industrial area. Watching from hides, Dr Kettlewell was able to see how the moths then fared. He witnessed redstarts, wagtails and other birds removing the moths and subsequent tests proved beyond doubt that each form was at a great disadvantage when on the 'wrong' background.

'Industrial melanism' was eventually found in more than a hundred other moth species living near dirty industrial centres of Europe and the U.S.A. Soot-adapted spiders and caterpillars were also located. Now that air pollution (at least with soot) is rapidly becoming a thing of the past, we may expect that the melanic varieties will again become the rarities they used to be. The trend has already begun.

Dead leaves on the forest floor create another situation where background picturing provides effective camouflage. The typical pattern is a bold patchwork of pale-brown, dark-brown and black, simulating curled dead leaves and their shadows. Examples may be drawn from almost every group of ground-dwelling animals, but snakes like the gaboon viper, *Bitis gabonica*, and puff adder, *B. arietans*, are spectacular because of the danger inherent in their camouflage. Pheasants, woodcock and other ground-nesting birds of wooded habitats also typify this kind of pattern. Noteworthy among the insects with dead-leaves patterning are numerous forest butterflies that become invisible the moment they land in the leaf-litter.

'Adventitious colour resemblance' is the term used for animals that use extrinsic objects for their camouflage. Some beetles coat themselves with chalk, sand or mud to blend perfectly with

Left above: The irregular longitudinal stripes of the *Chiromantis* tree-frog from Tanzania simulate fissures in the bark.

Left below: Two specimens of a noctuid moth at rest on lichen-covered bark in montane forest, East Africa. Individual variation in this species makes it difficult for insectivorous birds to form a 'search-image'.

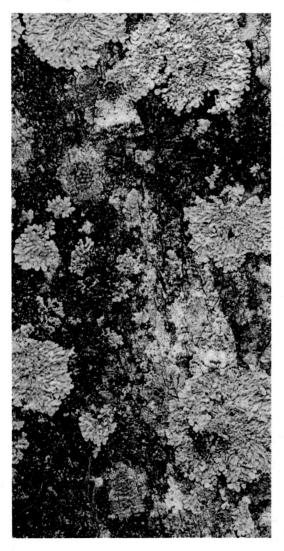

the substrate, while at least one species of looper caterpillar found in eastern Africa covers itself with lichen. The caterpillar's almost naked body bears a few stiff hairs to which it fastens bits of lichen until it is completely covered. Since it feeds on the same lichens, it is almost impossible to detect except when accidentally knocked from its resting place. The case-bearing aquatic larvae of caddis-flies can also be classified under this heading, especially those that build their cases of sand or mud particles. Many other animals attach objects to themselves to form a mask, but since the shape of the animal is altered as well as its colour they fit better in the later section on 'adventitious disguise'.

Most camouflaged animals have evolved a coloration that is the best compromise in their surroundings. It suits most of the situations the individual is likely to be in, but some better than others. A few species have improved on this position by developing the ability to adjust their coloration to that of the background – some slowly, others quite rapidly.

Slow colour adjustment, taking days or weeks to complete, is exemplified by many spiders, grasshoppers and caterpillars. The common African bark spider, *Hersilia*, occurs in several different colour forms even within a small area. Each colour form may be found resting on an Acacia, or other tree, the bark of which it matches exactly. Some individuals are brown, some dark-green and others a bright rusty-red, yet all of them may have hatched from the same batch of eggs for each individual can assume the colour of the tree on which it settles down after dispersing as a baby. Many other spiders undergo similar colour changes. Experiments with crab-spiders of various species have shown that the change performed when a spider is transferred, say, from a yellow buttercup to a white violet, is reversible when it is returned to its original flower.

Fast colour-change is generally associated with the Old World chameleons. The New World anole lizards are erroneously called chameleons by some people, but their colour-

changing ability is greatly inferior. Among true chameleons the colour range varies considerably between species. The East African flap-necked chameleon, *Chameleo dilepsis*, which lives among leafy branches, goes from dark-green with black markings to a uniform pale-green. Another species from the same region, Fischer's chameleon, *C. fischeri*, adjusts mainly between light- and dark-brown as befits an animal that spends most of its time on tree bark. Some individuals manage to develop a green tinge when resting in the foliage.

A number of marine animals also can perform rapid colour change. Flatfish, such as the plaice and flounder, are able to blend with a variety of sea-bottom substrates, ranging from pure sand to coarse gravel. As a plaice settles on the bottom and adjusts its pattern it performs a wriggle, sending up a flurry of sediment that settles over the margins of the body. This simple action serves to obscure the symmetry of the fish which might otherwise spoil its camouflage.

Surprisingly, the fastest colour adjustments of all are made by molluscs. Many squids,

Right: an example of 'adventitious colour resemblance'. An African looper caterpillar that fastens lichen to stiff hairs on its near-naked body.

Below: One of the many anole lizards from tropical America. The strongly disruptive pattern is well suited to the broken background of the jungle floor. Like the Old World chameleons, many anoles can adjust their body colour quite rapidly.

cuttlefish and octopuses can change colour instantly as they cross different backgrounds. Moreover, some octopuses can even alter their skin texture to imitate that of surfaces over which they glide. In their social life also, and particularly in encounters between the sexes or rival territory-holders, they go through striking colour-changes.

Apart from these impromptu adjustments, there are colour-changes that occur on a regular basis: fishes and frogs are often coloured quite differently at night from what they are by day. Night colours are also adopted by some reptiles. Baby Fischer's chameleons, which are almost impossible to locate during the day, can be found with ease at night by the light of a torch. Before settling down to sleep these tiny reptiles crawl to the ends of thin twigs where they curl their tails in a tight coil. Their colour becomes a pale purplish-pink and by torchlight they are easy to see against the green vegeta-

tion. But for a nocturnal predator with acute but monochrome vision, they must be hard to distinguish from the pale leaves. Brightly coloured coral fishes change in the opposite direction, becoming more sombre at night. Presumably, this colour-change is adaptive, but at present too little is known of fishes' behaviour at night to allow any definite explanation.

There is no difficulty, however, in explaining the value of the seasonal colour-changes undergone by Arctic birds and mammals. In the short summer, the majority lose their white coat and become dark. Exceptionally, the polar bear, 'polar' hare (the northernmost race of the Arctic hare), snowy owl, and the white phase of the gyrfalcon (the so-called Greenland falcon) remain permanently white. Polar bears are essentially marine animals, spending most of their time searching ice-floes for seals, their main prey. For them, white provides good

Above: Crab spider lying in wait for its insect prey. Slow colour adjustment allows these spiders to move between flowers of different hues; the camouflage probably serves mainly for protection from birds. A miniscule male is attached to the female's abdomen.

camouflage throughout the year. The polar hare would seem to have an insuperable problem in summer, but it outwits its enemies by feeding close to the edges of the snow patches that persist through the warm months. The snowy owl and Greenland falcon cannot avoid becoming conspicuous when the snow melts, and this would create problems in their hunting were this not the time of year when their food is most plentiful. Their summer whiteness is a real liability only when nesting. They are able to defend the nest against small predators like stoats, and if a more powerful enemy approaches the incubating parent slips away, leaving the eggs or young protected by their excellent camouflage.

Most of the other Arctic breeding birds fly south for the winter, but a handful of species, able to find food even during the coldest months, stay permanently. Of these, the snow buntings, rock ptarmigan and willow ptarmigan moult into a white plumage as winter time ap-

proaches, but the raven remains forever black. Like the desert ravens mentioned earlier, they have no need of camouflage, neither for defence nor for hunting, and therefore retain the colour that makes them most visible to other ravens.

Most of the Arctic mammals remain in the far north throughout the winter and a number change to a white coat. At high latitudes practically all the foxes, stoats, weasels and Arctic hares become white in winter. Even the reindeer living at the northern edge of their range become almost white, though further south they do not change colour when moulting into the thicker winter coat.

In other habitats, seasonal colour-change for purely camouflage purposes is unusual. One instance is provided by the sika deer of Japan, which is spotted in summer to suit the dappled sunlight in deciduous forest but plainly coloured in winter when the light is more uniform. To a lesser degree the European fallow deer makes the same seasonal change.

Left: Octopuses can change their skin texture as well as colour to conform with that of the surroundings.

Above: Creatures as large as the giraffe can benefit from disruptive patterning.

Background colour-matching alone provides concealment, but it does not hide the body outline, and in some situations this can be a serious shortcoming. The characteristic shape of a fish, frog or other animal may give it away even though it matches the surroundings in colour. This is especially so when the background consists of regularly shaped objects like pebbles or leaves. In such situations, 'disruptive patterning' serves to break the outline and make detection by predators difficult.

Paradoxically, an outline is best interrupted by the incorporation of strong-coloured areas into the pattern. These areas stand out from the rest of the body which is coloured to match the background and draws the eye towards their bold, but meaningless shapes.

Seen out of context in a zoo or museum, animals bearing strong disruptive patterns may look as if they would be conspicuous anywhere.

But in their natural surroundings they prove to be exceedingly cryptic. A giraffe, for instance, might seem pre-eminently unsuited to merging with any background, yet in the wild they can prove very difficult to locate because of their disruptive pattern. They are probably most at risk from attack by lions when feeding intently among tall thorn-trees, especially at night. Here, moving with slow deliberation among slim acacias, the ungainly-looking animals blend surprisingly well with their surroundings. The long legs and neck disappear among the slender branches, and the short stubby horns — so useless for fighting — simulate the stubs of broken branches. The disruptive pattern of angular spots is particularly effective as the animal stands amid the finely-leaved trees through which dappled light penetrates to ground level. Admittedly, giraffes are blatantly obvious when out in the open, but they are not then vulnerable. From such a high vantage point approaching lions are easily spotted and avoided, for giraffes are swift runners.

Many ground-nesting birds (pheasants, ducks, sandpipers, etc.) have a variegated pattern that combines background picturing with disruptive elements. The disruptive effect is most pronounced in plovers belonging to the genus *Charadrius*, which have bold stripes on the head and prominent black collars. The bird's brown back blends with the background, while the black-and-white pattern of the head and breast creates a number of conspicuous but irrelevent shapes. Clearly these shapes must not stand out too prominently, if they are not to be counter-productive by drawing attention. How bold they can be, and remain effective, depends on the surroundings. The European ring plover, *Charadrius hiaticula*, which nests among pebbles, has strong disruptive markings, whereas the closely related Kentish plover, *C. alexandrinus*, which favours sandy places, has a less prominent pattern, creating shapes no more obvious than the odd sea-shells scattered about the sand. A similar distinction may be made among New World species by comparing the boldly-marked kildeer plover, *C. vociferus*, with the piping plover, *C. melodus*.

Bold, dark stripes or spots have a strong disruptive effect on the appearance of many species. At the same time they can be regarded as a constituent of background-matching in that they simulate shadows. Clearly it is unwise to make too great a distinction between the two principles. Animals hiding among straight shadows tend to have black stripes. The tiger is a familiar example, but reed-dwelling birds, frogs, grasshoppers and other small animals have comparable parallel striping. In the resting position the stripes continue unbroken across different parts of the body and folded limbs, a

Below: The American kildeer plover has a strongly disruptive pattern breaking its outline into a number of shapes that are meaningless to predators.

feature sometimes referred to as 'coincident disruptive patterning'.

In dense forest, where shadows are broken and irregular, the black, disruptive markings on camouflaged animals tend to be spots or small geometric shapes. Ocelot, jaguar and leopard show this well, as do many snakes. Indeed, it is only the strongly contrasted disruptive patterns that make the skins of forest cats, pythons and boas so highly prized as human adornments.

Another important principle of camouflage is countershading. This is a method for overcoming the conspicuousness that the rounded solidarity of an animal would otherwise produce in situations where there are no other rounded objects. Without countershading a globular spider might fit well into a background of pebbles, but among flat leaves, or on the bare ground, its rounded appearance would be a give-away. Countershading should be regarded therefore as optical flattening, serving the same purpose as true body flattening which is possible for relatively few species. The principle is simple: with the light coming from above, as it usually does, the back of an animal is better illuminated than its underside. If it were of uniform tone all over, the back would therefore appear much lighter than the belly. This difference in tone, and the transition from light to dark on the flank, would show it to be solid and rounded. It is for this reason that in well-lit situations camouflaged animals are not uniformly toned. The back, which receives most light, is darker than the underside, and a gradation from dark to light on the animal's sides compensates for the fall-off in illumination there. The apparent uniformity, which is normally associated with flatness, tricks the brain of the observer into thinking that the object is indeed flat.

Countershading features in the camouflage of most vertebrates. In fishes, frogs, reptiles and mammals it is generally quite obvious, but with birds the situation is often complicated by patches of bright colour. Spiders, caterpillars and other fat-bodied invertebrates living in brightly-lit situations also exhibit typical countershading. Should there be any doubt that the countershading has been evolved for optical flattening, one has only to look at those animals that live permanently upside-down. Caterpillars, like that of the oleander hawk-moth, *Daphnis nerii*, the 'upside-down catfish', sloths and other such creatures all show reversed (or inverted) countershading, that is, they are darker ventrally.

Among land animals real flatness has been evolved mainly by smaller species that live on rock-faces or the smooth bark characteristic of many trees, especially in tropical forests. Bark-spiders in the genus *Hersilia* are extremely flat and normally rest throughout the day spread-

Above: Caterpillar of the copper underwing moth in Britain. Like other animals that rest upside down, it achieves optical flattening by having a pale back and a dark belly — the opposite to normal countershading.

eagled and quite still. Catching them can be quite an art, for as one approaches the tree, spiders resting on that side are prone to dash round to the other side. The trick is to pick a slim tree, then to stand still and pass your hands up and down the far side. The bark-spiders, and often mantids, too, then come racing round to flatten themselves on the side you are watching. Other animals that have become flattened for resting on tree-trunks include cockroaches and gecko lizards. All of them become active at night when they move about with ease, their bodies raised above the surface on surprisingly long legs.

On the large leaves of tropical forest trees, a different range of flat creatures, including grass-hoppers, spiders and tree-frogs may be found. The New World *Phylomedusa* frogs even have flattened skulls to aid their crypsis.

Not many nocturnally-active animals rest immobile on the ground by day — there would be a considerable risk of being walked on if they did — but some day-active animals have a squat form that helps their camouflage when they freeze in times of danger. Horned toads, *Phryno-soma*, of the American deserts are in this cate-gory; when disturbed they wriggle sideways till partly covered with sand.

No birds or mammals have acquired a truly flattened body shape, but many flatten them-selves as best they can when in danger. The chicks of plovers and other ground-birds press themselves hard against the ground, neck out-stretched, when they hear the parents' alarm call. Their disruptive pattern renders them in-conspicuous from the air, while the low profile reduces their visibility to ground-predators.

In the sea there are probably far more flat animals than there are on land. Mainly this is because locomotion in water is less hampered by

a flattened body-form. Indeed it is well-suited to certain modes of swimming as shown by the giant manta ray which is certainly not flattened for concealment. A second reason is that animals can rest on the sea-bottom for long periods with no risk of being walked upon. There are more than six hundred species in the main family of flatfish, which includes the plaice, flounder, dab and sole. Largest is the Atlantic halibut, which can weigh as much as 700 lb. (320 kg.).

The patterned upper surface of a flatfish is actually one side of the animal (the right side in some species, the left in others). The young flatfish starts life shaped like other fishes, but as it matures its body becomes compressed from side to side. At the same time the eye on the side destined to be underneath moves over the head to the other side, so that binocular vision is retained. In its final form a flatfish is white or silver beneath, which provides camouflage against the silvery sea surface when it is swim-ming off the bottom and vulnerable to attack from below.

No matter how good the overall camouflage of an animal, it would be rendered useless if some conspicuous feature were left to reveal its presence. For vertebrates the eye poses a special problem, since it is an intrinsically conspicuous object. For good optical performance it must be large and round, with a shiny convex surface. At its centre is the pupil which always appears black since it is a hole (through which the light passes to the retina). Owls, nightjars and other nocturnal birds that rely on camouflage by day can hide their eyes easily. They just close the lids which bear the same cryptic pattern as the rest of the head. Actually they do not fully close their eyes: a chink is left through which

the resting bird can see enough of what is going on for its safety.

Fishes, amphibians and reptiles cannot cover the eye this way since they lack eyelids. Instead, they have developed special eye-camouflage: a bold black line passes through the centre of the eye, as part of the general disruptive pattern, and the black pupil disappears as part of it. The eye-stripe is usually horizontal, but in some species it runs vertically through the pupil. Alternatively, there is no eye-stripe and the pupil is reduced in size by numerous tiny muscles surrounding it. In some geckoes it

Above: Morphological flattening is best shown by flatfishes such as the plaice. When settling on the sea-bed, it adjusts its colour to match the background, and flicks sand over its body margins.

Left below: Warned of danger by its parents' alarm calls the courser chick flattens itself on the ground and remains motionless.

becomes a chain of small holes just large enough to permit a constant watch.

Long limbs or a tail sticking out from the body may also reveal an animal's presence. For this reason camouflaged animals usually tuck their limbs under the body, or fold them neatly along the sides. If there is a strong disruptive pattern on the body it is coincident across the folded limbs, helping to mask their form.

Coincident disruptive patterns are a notable feature of the camouflage patterns of frogs and toads because of their long legs. The disruptive patterns on many moths run without a break across both pairs of wings, but coincidence can be seen only when the animal is resting normally. Pinned specimens often have the wings and legs splayed out in an unnatural way and the various sections of the stripes do not meet up properly.

Animals resting in well-lit situations can be let down by their own shadow, no matter how well their coloration matches the background. We have already seen how a low profile mini-

mizes the length of the shadow as well as reducing the contours, but there are other adaptations aimed at shadow reduction. The commonest is a marginal fringe or flap. Many of the Lasiocampid moth caterpillars that rest by day on tree-trunks have stiff hairs around the body. They fill the space between the sides of the body and the surface, effectively smoothing over the caterpillar's contour and reducing its shadow. Some arboreal geckoes and lizards have projecting flanges of skin round the body and tail serving the same purpose. Apart from any other advantages, the habit of crouching when in danger, shown by many birds and mammals, helps to reduce the shadow. Similarly, some of the butterflies that rest on the ground in sunlit places tilt themselves sideways when they land.

Camouflage can be effective only when it is combined with stillness. As mentioned before, many cryptic species are active only at night. Of those active by day some walk so slowly that the motion is not perceptible at any distance. Chameleons have a particularly deliberate motion, advancing with a curious forwards-and-backwards rocking movement of the body. This makes their progression hard to detect among wind-disturbed foliage and shifting shadows.

Before it settles down for a long period a cryptic animal must ensure that it is on a suitable background. If, like many insects, it simply rests on the food plant it has little difficulty, but for many species a search has to be made. A moth, for example, that has been flying about during the night must find just the right kind of bark, or other special resting place before the arrival of dawn, and the early birds. The procedure has been studied experimentally with captive moths offered a choice of resting sites. Even within the same polymorphic species individual moths selected the background most appropriate to their coloration. Moreover, the final resting situation was governed partly by tactile as well as visual clues.

Birds that nest in situations where camouflage is to be their main defence for several weeks must obviously select the site with particular care. The Nubian bustard, which breeds along the southern edge of the Sahara, simply scrapes a shallow depression in the sand for its two eggs. With so much empty country to choose from it might seem that the nest-scrape could be almost anywhere. Yet it is generally in a very special place where the last remains of a fallen tree lie decaying on the ground. There, on the sand between the rotten branches, it lays its eggs. It is easy to see why it chooses such a spot. The bird's contour is far less obvious to jackals, and other ground predators hunting at night, than if it were sitting on uncluttered ground.

# Disguise

There can be no doubting the effectiveness of camouflage, but to most people disguise is even more wonderful. An animal shaped like a stick, stone or leaf seems somehow better adapted than one that is merely camouflaged. Yet there is no logical basis for such a notion. Indeed, one could well argue that some species have been able to derive sufficient protection from camouflage alone, while others have been obliged to go further and assume a disguise. This line of argument appears to imply that disguise is the inferior adaptation, but this is equally wrong. Each species has evolved the best system for its particular way of life. Whether an animal is camouflaged, disguised or for that matter warningly-coloured, depends on an impossibly complicated set of factors operating during its evolution. Always the result has been a compromise, but we may be sure it is the best possible compromise. Alternatives, however slightly inferior, have been removed by the relentless pressure of natural selection.

Although disguise might seem to confer only extra advantages, in reality there is also a cost. For disguise involves structural alterations to the ancestral body-shape, usually to the detriment of mobility. It is no coincidence that disguised animals are generally clumsy when compared with more normal-looking relatives. Mantids and grasshoppers which have be-

The European buff-tip moth shows near-perfect disguise as a broken twig. The twig's cross-section is imitated at both ends of the body, and the rolled wings give roundness. The moth finds its day-time resting place in the dark, possibly by its sense of touch.

come modified to look like flat, green leaves move awkwardly, and can become stuck in thick vegetation. The more usual slim-looking mantids and grasshoppers are positively nimble by comparison. Of course it is only the more extreme disguises such as these that slow an animal down. Minor changes in body-form like the two tufts of feathers on the head of a long-eared owl, part of its tree-stump disguise, are no encumbrance. The modification of the bird's shape is superficial and well away from the 'moving parts' — the neck, wings and legs. Similarly, the irregular outline of the wings of butterflies that look like dead leaves may not greatly impair their flight.

The most widespread disguises adopted by terrestrial animals are as fresh or dead leaves and sticks in wooded situations, as grass stems in semi-arid regions, or as stones in the desert. In the sea there are equivalent disguises, such as seaweed fronds, blades of eelgrass and lumps of broken coral. Less well known are the more subtle disguises, like plant seeds, thorns, flowers and even the shrivelled stigma on a ripening fruit. Animal faeces, especially bird droppings, have also been copied. They are, after all, common enough objects and certainly of no interest to predators.

The most convincing examples of green-leaf disguise are seen in the Indian leaf-insect, *Phyllium*, various katydids (or long-horned grasshoppers) and a strange mantis, *Choeradodis rhomboidea*, from tropical America. *Phyllium* has large flat wings, shaped and patterned exactly like leaves. Wide flanges on the legs and body simulate parts of more leaves, so the insect resembles a whole bunch of them. In the katydids the illusion is given almost entirely by the large fore-wings which are held against the

sides of the body so that they project like a high pitched roof above the back. All of them have details of leaf venation traced on the 'leaf' surfaces. Several katydids have improved the leaf imitation further by having irregularities along the wing margins imitating places where insects have been chewing, and pale spots like those made by moulds on a real leaf. In the tropical rainforest where they live, new leaves, unblemished by insect or fungal attack, are few and far between so insects are more inconspicuous if they resemble old leaves.

These insects have the flatness required by leaf disguise but the same deception is possible also for animals with a rounded shape. A small chameleon, *Rhampholeon boulengeri*, found in the Congo forests is coloured green like the leaves among which it lives. Its pointed head simulates a leaf-tip and its short, stiff tail (unusual for a chameleon) looks like the stalk. Though the chameleon's body is as compressed as it can be, the look of extreme thinness is mainly illusory, being produced by countershading.

Strange as it may seem, a far greater variety of animals have evolved disguise as fallen leaves than have copied green ones. Speculating a little, one can readily suggest a reason for this: leaves on a plant are all fairly uniform, but they can differ considerably in shape and tone between species. This means that an animal disguised as a particular leaf must always live among the leaves it imitates, which may be simple enough for specialized herbivores like the Indian leaf-insect, and leaf-like katydids which resemble their food plant, but creates problems for animals who range widely. The mantid, *Choeradodis*, mentioned earlier as an extreme example of leaf-disguise, resembles the circular leaf (or, rather, two leaves) of a climbing plant. This particular climber grows commonly up the trunks of forest trees in the Central American jungle, so by resting on any tree trunk by day the mantid is likely to go unnoticed. Though not directly interested in the leaves it copies, it may be able to find them, using visual or olfactory cues, before settling down after a night's hunting.

Dead leaves on the other hand are far less uniform. Leaves of various shapes get mixed together when they fall, and become even less distinctive as they change colour, curl, and decay. Accordingly we find a wide range of

Above: A 'classic' disguise from tropical America. The preying mantis *Choeradodis rhomboidea* rests during the day near creepers whose leaves it copies perfectly.

Right: The European brimstone butterfly needs its convincing leaf disguise on cool days and when hibernating. On sunny days it can escape from most enemies by fast flight.

Below: A horned frog *Megophrys nasuta* from S.E. Asia, disguised among dead leaves. Lateral flaps on the head, like pointed leaf-tips, appear to cast shadows which are, in fact, the huge eyes.

animals displaying dead-leaf disguise. Apart from the expected insects, there are spiders, frogs, lizards, and even fishes.

It might be thought that on land the only place for an animal looking like a fallen leaf is on the ground. This may well be the case in many areas, but inside the lowland jungle, where there is scarcely ever a breath of wind, falling leaves tend to lodge in all kinds of precarious situations. They get tangled in vines and epiphytes, come to rest on large living leaves and get caught in spiders' webs where they stay for weeks if no animal disturbs them. This is why animals disguised as dead leaves are to be found at all levels in the jungle. Furthermore there are many plants that retain their old leaves long after they have turned brown — providing yet another opportunity for this form of disguise.

The horned frog, *Megophrys nasuta*, of Malaya, is one of the classic examples. As befits an animal of the jungle leaf-litter, it is patterned in subdued browns. Ridges along its back simulate the midribs of fallen leaves and dark tubercles look like small galls. The most curious features, however, are the pointed flaps projecting one above each eye. These continue the flat-looking back of the animal, bringing it to two sharp points to give the appearance of overlapping leaves. The huge eyes are hidden in the black patches that imitate broad shadows under the flaps of skin. A toad, *Bufo superciliaris*, living in the same habitat, but in West Africa, has an almost identical disguise, with the same pointed flaps overshadowing the eyes — a striking example of convergence in evolution.

Equally 'classic' are the leaf butterflies, *Kallima*, of the Old World tropics. On the upperside they are brightly coloured, but underneath is a perfect representation of a dead leaf. Both fore- and hind-wings terminate in a point so that the shape of the butterfly at rest is precisely that of a lanceolate leaf. The blotched pattern of browns give the look of decay while strong brown lines running from point to point represent the leaf's main vein. Throughout the world other jungle butterflies which rest on the forest floor have evolved the leaf pattern, but none so precisely as *Kallima*. There are also legions of moths in this category. Many rest on the ground, but some, like the lappet, angle-shades and eyed hawk-moth, to name but three European species, rest among foliage. The angle-shades rests with its wings crinkled, as if deformed, imitating a shrivelled leaf.

Broken bits of dead leaf are copied by small moths, beetles and other insects. Spiders, too, have adopted this disguise. Fragments of dead leaves often get blown or fall into their webs, so, disguised as a fragment of leaf, a spider can sit in full view without much risk of being noticed. One orb-web spider from Costa Rica, though not itself protected by this disguise, uses it to protect its eggs. Just before laying, the female constructs a simple and rather untidy-looking web to hold the flat, brown egg case which is left to look like a piece of leaf hanging in an old, abandoned web.

The leaves which fall into forest streams and float downstream also have their imitators. The South American leaf-fish, *Monocirrhus polyacanthus*, is a predator and grabs at any unsuspecting small fish that come too near. It hangs motionless in the water with its head down, moving along with the real leaves it so closely resembles. Another fish, *Platax pinnatus*, of the Indian Ocean belongs to the family of bat-fishes. The young *Platax* lives in shallow waters, especially in mangrove-lined creeks. Normally it swims about actively, but if an egret wades near, or a tern passes overhead, it suddenly becomes listless. Its laterally-compressed body flops around in the waves just like one of the many mangrove-tree leaves floating in the water. The leaf shape of the little fish's body is not spoiled by its fan-shaped tail which is completely transparent. The eyes, which might also reveal its true identity, are concealed by prominent eye-stripes running through the pupils.

A special resemblance to sticks has been evolved by a wide range of insects, including grasshoppers, praying mantises and, of course, stick-insects (phasmids). This disguise requires an elongate body, with little or no separation between head, thorax and abdomen. The body and legs are generally mottled grey or brown and covered with minor irregularities. Wings, if any, are kept tightly wrapped around the abdomen and are difficult to see.

Some 2,000 species of stick-insects (or walking-sticks) inhabit the tropics. They vary in size from small, delicate insects resembling grass stems to monsters a foot in length, looking more like small branches. Some can adjust their colour to suit the surroundings, and all are slow, deliberate movers. Not all phasmids look like sticks, however. An oddity from New Guinea, *Extatosoma tiararum*, begins life as an ant mimic, racing about madly with its front legs flaying just like antennae. After a while it settles down, resuming the more typical appearance of its group – a pale stick, but when threatened it is able to transform itself into what looks like an angry scorpion!

Left: A young bat-fish *Platax* rests on its side among mangrove leaves in a Kenyan estuary. The deception created by a leaf-like body is aided by a scarcely visible tail, while the eye is obliterated by an eye-stripe.

night this oddly-shaped arachnid, bearing two long tubercles on its back, sits on a large orb-web constructed among the branches. But by day, neither the spider not its web are to be seen. Just before dawn, the spider selects a resting place on one of the thorny branches nearby, presses its body tightly round the stem and tucks its legs well in. It then looks like a pair of broken spines arising from an irregularity on the branch. As for the web, this is eaten by the spider before it settles down. Were it to be left to be used again it might lead a diligent bird to the spider resting nearby.

Among vertebrates, disguise as thin stems and twigs is confined mainly to arboreal snakes. Snakes are obviously pre-adapted to twig disguise, due to their elongate, cylindrical body form. But the simulation is improved in certain slender tree snakes which have long snouts instead of the usual blunt head. The long-nosed tree-snake, *Dryophis nasuta*, and the Madagascan rear-fanged snake, *Langaha intermedia*, exemplify this adaptation.

Several birds imitate larger branches most convincingly: they are all superficially similar, though belonging to different families. These rather bizarre forms are included in the Capri-

Left: Stick insects are plentiful in the tropics. They rest immobile by day, but become quite active at night.

Below: This African spider remains motionless during daylight hours, relying on its disguise as thorns. The huge orb-web, which it spins every night, is consumed before dawn so as not to reveal its whereabouts.

Among the moths that imitate broken twigs, none is more remarkable than the buff-tip, *Phalera bucephala*, a common European species. It is frequently portrayed wrongly in diagrams and artificially-posed photographs by being placed on a tree and made to look like a broken side-branch. If this were the buff-tip's disguise it would be hard to explain why it has the pattern of a broken twig at both ends. At its front there is the perfect copy of a twig's cross-section on the head and thorax, and at the back is a similar pattern borne by the juxtaposed wing-tips. In reality, the buff-tip is disguised as a short, snapped-off bit of twig which gives it protection as it rests on the ground.

Insects that truly resemble side-branches rest with their body, or part of it, sticking out prominently from the main stem. Many of the caterpillars in the family Geometridae — called 'loopers' in Britain and 'measuring worms' in the United States — stiffen quickly when disturbed, like a broken side-branch. This pose can be maintained for hours if necessary.

In tropical forests many branches, palm stems and lianas are protected from browsers by thorns or stinging hairs. Here and there among the tangles a perfectly innocuous vine with a smooth surface looks as if it is covered with thorns. In fact, the thorns are plant-bugs feeding in line on the vine's sap. Similar thorn-like tree-hoppers (or membracids) can be found throughout the tropics.

The African spider, *Caerostris*, provides another good example of thorn-disguise. By

Above: Fischer's chameleon is an arboreal species from Northern Tanzania. The two serrated 'horns' simulate the outline of broken bark.

mulgidae (nightjars, nighthawks and poor-wills) with a world-wide distribution; Nyctibiidae (potoos) of tropical America; and Podargidae (frogmouths) of Australia and south-east Asia. All feed at night on flying insects, hence their enormously wide beaks and rictal bristles. Furthermore, they all rest out in the open by day. This last fact explains their similar patterning – a combination of irregular dark markings and variegated brown speckles – simulating rough bark.

The nightjars and frogmouths tend to rest in a horizontal position. Some species favour the ground under trees, where they resemble pieces of wood; others prefer a horizontal limb where they can look like a broken side-branch. The potoos differ in that they habitually rest at the tip of a vertical stump, the tail being pressed firmly against the bark to hide the discontinuity between bird and branch. If it senses danger the potoo points its bill skywards and holds this stiff position as long as necessary. The breeding habits of the potoo are also of interest. The single egg is wedged in a crevice and kept covered constantly by one of the parents. The young potoo remains alone for long periods, protected by a dead-branch disguise every bit as good as its parents'. Many owls also adopt a stiff, upright posture during the day, in imitation of a stump. Their patterning is reminiscent of the nightjars etc. described above.

Also, like them they sit with their eyes almost closed, though they can see well enough through the remaining slit. Many of the owls with this type of disguise have tufts of feathers on the head from which common names like horned owl and long-eared owl derive. Presumably the deception is not directed at the more dangerous enemies but at small birds which will incessantly mob any owl they detect.

The largest creatures in the category of dead-wood disguise are the caimans, alligators and crocodiles. Large specimens have been reliably described as looking like floating logs as they swim slowly or drift towards unsuspecting animals drinking at the water's edge. It is conceivable that very young crocodiles benefit from the disguise by escaping the attention of avian predators.

Special resemblance to flowers for defensive purposes would seem to be fraught with problems. Apart from the high degree of adaptation required, flowers tend to be conspicuous. Indeed they have been evolved expressly to attract visitors. This means that any creature disguised as a flower is liable to frequent disturbance. And disturbance could lead to movement which might be detected by a watching predator.

Bugs in the genus *Ityraea*, confined to tropical Africa, are brightly coloured, moth-like insects. They feed gregariously on the succulent stems

Above: The South American pottoo feeds on insects during the night. By day it perches among the branches, assuming a stiff vertical posture like a broken stump whenever it senses danger.

Overleaf: Thorn tree-hoppers (mebracids) on a jungle liana, C. America. The adults arranged on top are disguised as green thorns, while the nymphs below resemble rough bark.

of herbs, arranging themselves in lines near the tip of a vertical shoot. The effect is rather like an unopened lupin flower. If it is disturbed, however, the inflorescence explodes as the bugs fly off erratically in all directions. There are also so-called flower-mantids such as *Hymenopus* in South-east Asia and *Pseudocreobotra* in Africa. By virtue of their white or pink coloration and petal-like form, they are able to rest undetected among flowers of similar hue. Their prey comprise visiting bees, butterflies and other insects. Flower-mantises are usually described as having 'alluring coloration'. The implication is that insects are lured to what they mistake for a flower, only to be seized and eaten. Certainly insects searching flowers for nectar might well visit a waiting flower-mantis, but it seems improbable that the mantis needs the disguise to attract or avoid detection by its victims. As long as it remains motionless on a flower it would probably not be noticed by the flower's visitors whatever its colour. The disguise probably serves mainly to protect the mantis from insectivorous birds. After all, mantises belong to a palatable group of insects, and by lying in wait for flower-visiting insects they put themselves in a very prominent situation.

Brightly coloured fruits might seem to offer good opportunities for convincing disguise, yet few species have taken to this particular resemblance, and the reason is not hard to

find — such fruits are appealing to birds and palatable insects looking like a ripe fruit would therefore risk being eaten by frugivorous birds.

There is no such danger in copying unripe fruit, and many insects and spiders do just this. In East Africa, one minuscule caterpillar rests by day at the tip of the unripe green berry on which it feeds at night. Its stiff, contorted body replicates the shrivelled floral parts still attached to many berries at this stage of maturation. Caterpillars of the European wood white butterfly imitate the seed pods of the food-plant, having precisely the same green colour, shape and texture as the pod.

The black-and-white droppings of birds are a conspicuous feature of most habitats. They are of no interest to the birds themselves. Consequently a wide variety of insects and spiders have evolved extremely convincing disguise as excrement. It is so convincing, in fact, that to find such insects one has to look closely at every dropping and white splash. Most prove to be what they seem, but here and there will be found some incredible creatures. Most of them are unwilling to give themselves away until touched. Even then, those that resemble dry droppings may allow themselves to fall to the ground. An inexperienced naturalist can easily be taken in by this little trick.

Moths disguised this way are coloured white with black markings concentrated at one end

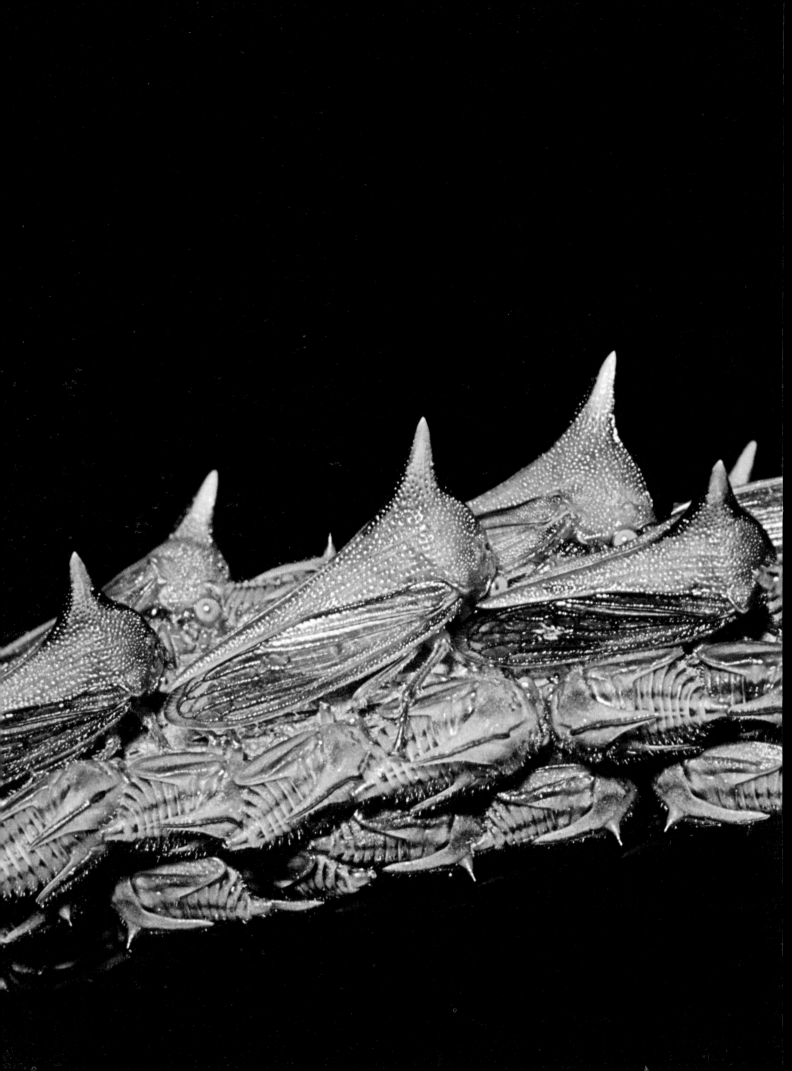

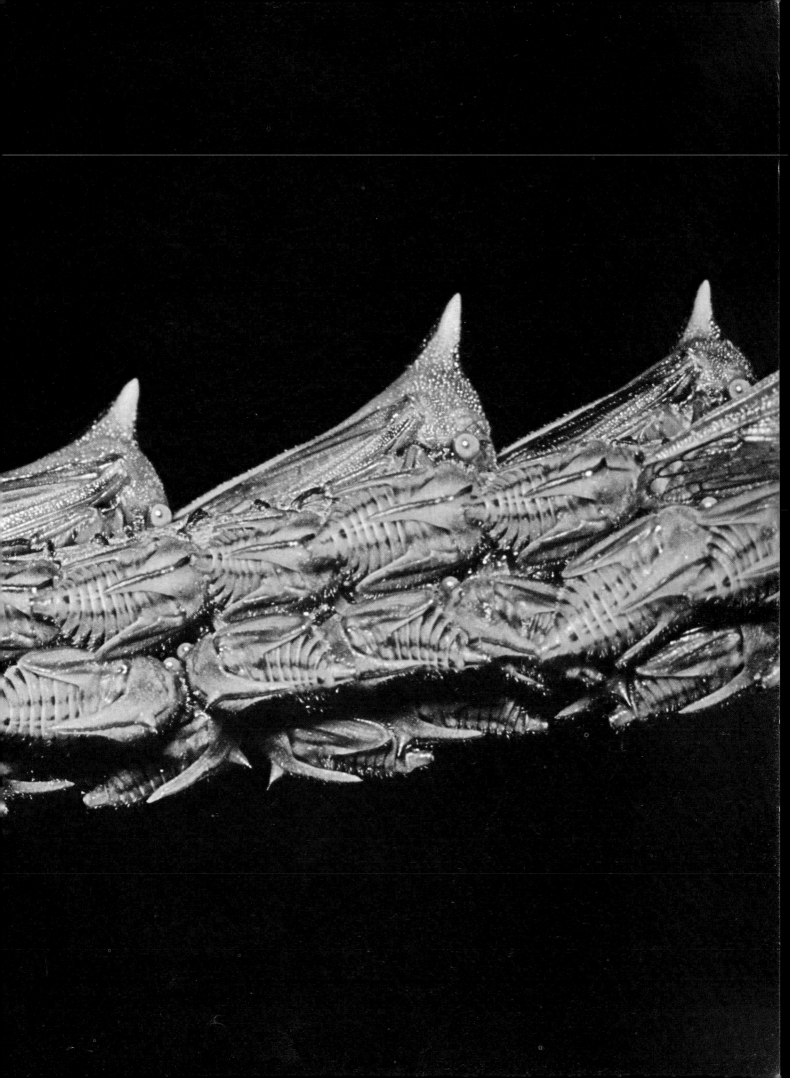

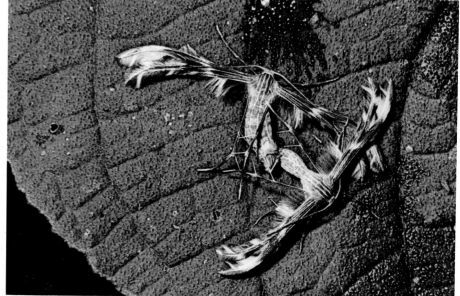

of the body. Among European species the lime-speck pug, *Eupithecia centauriata*, and *Cilix glaucata*, known as the Chinese character, sit with wings rolled along the body to resemble dry droppings. The clouded silver differs in resting on top of a leaf with wings outstretched and looking like the white splash produced by a wet bird-dropping. Other striking European examples are the caterpillar of the alder moth, *Apatele alni*, and chrysalis of the black hairstreak butterfly.

In the tropics there are even more fantastic examples: the caterpillar of one Central American hawkmoth is a perfect copy of the dropping of a bird that has been feeding on berries and has numerous small seeds in its faeces. It sits all day in the middle of a leaf, on which it feeds at night, curled round so that flaps on either side of the head overlap the tail. The 'seedy' look is produced by pale tubercles on its brown skin.

It would be an unending task to list all the strange disguises adopted by animals, even if they were all known, but a few more can be mentioned to illustrate the great diversity.

Grasshoppers in desert regions are extremely vulnerable to birds. Some are protected by background colour-matching alone; others are disguised as bits of the dry grass stems that sustain them, or as small stones. *Eremocharis insignis* is a small species living in the rocky deserts of North America. When alarmed it tucks its legs and antennae snugly against the head and body to smooth its contours. Combined with the overall buff colour and finely-pitted texture of the body, this behaviour completes the simulation of a small, weathered stone-fragment lying among the shattered and sand-blasted rubble.

Larger stones, weathered by water action

into smooth pebbles, have been imitated by the eggs of plovers and other birds nesting on shingle banks. Birds' eggs have an intrinsic pebble-shape, so the only special requirements are appropriate colouring, and absence of any obvious nest material. One give-away feature (at close quarters) might be the identical coloration of two or three 'pebbles' lying together. This is usually avoided by the eggs in a clutch differing from one another in their markings. The eggs of the stone curlew and ringed plover are so well protected that when danger threatens the incubating parent (though beautifully camouflaged) slips away. The departure is made with great care. The bird creeps quietly across the pebble bed and then flies off conspicuously from a point many yards from the nest.

Searching for the nest near the place where the bird took off is a waste of time, and a wider search involving close scrutiny of all the pebbles on the beach is a daunting task indeed. The human observer can eventually find the nest, on a second or third visit, the position being revealed as one watches the plover leave the eggs when a second person walks towards the nest area. A fox or other predator can only search around over a large area and needs much luck or patience if it is to find the eggs.

The behaviour of the Kentish plover in this respect is particularly interesting. Unlike the ring plover it nests on sand at the top of the beach where there are only sea-shells or a few scattered pebbles. Before creeping away from the nest, the parent kicks sand over the eggs, partly covering them. A different proportion of each egg is left showing and instead of three or four similar-sized 'pebbles', a marauder sees only a group of differently sized objects, as could occur naturally.

Above: Two striking examples of disguise from the forests of Costa Rica. The hawk-moth caterpillar (far left) imitates the seed-packed excrement of a frugivorous bird, while the pair of plume-moths simulate a feather stuck to a wet leaf.

A feather might seem an unlikely model for disguise, but there is one plume-moth living in the Costa Rican cloud-forest that looks exactly like a soggy white feather stuck to a leaf. The deception is convincing because a real down-feather would look just the same in this wet habitat. It was encountered in the month of January, when very few of the birds in this region breed. Presumably it would not be on the wing later in the year when birds are seeking small feathers to line their nests!

In the sea, opportunities for disguise and the need for this form of protection are equally as great as on land. Unfortunately, knowledge of the undersea world is still so meagre that we can only guess at the adaptive value of the strange shapes of many fishes, molluscs and other marine organisms. Nevertheless there are some easily understood examples.

The principles of disguise are the same as in terrestrial habitats, but the models are quite different. Predatory fishes are the principal enemies, though there are also sea-snakes, dolphins and seals; squids and octopuses have acute vision, too.

Surface-living creatures are predated by terns, pelicans and other sea-birds. Along the shore and in coral pools, reef herons, king-fishers and waders create additional dangers for palatable fishes and invertebrates.

Disguise as sea-weed has evolved separately in several families of fishes. Among the masses of floating 'gulfweed', in that part of the Atlantic known as the Sargasso sea, lives the Sargassum fish, *Histrio*. It is coloured exactly like the weed and papillae on its skin simulate the crenulated edge of the fronds. Several sea-horses, *Hippocampus*, are also associated with sea-weeds and have a 'mane' of frondose appendages along the back as part of their disguise. The related sea-dragon, *Phycodurus eques*, has even longer branched appendages all over its body.

Below: An East African shrimp which has 'planted' itself with pieces of living algae, an example of adventitious disguise.

Razorfishes, *Centriscus scutatus*, take their name from their long, parallel-sided and extremely compressed form. They are patterned and shaped like blades of the eel-grasses among which they find food. An unusual swimming posture, with the pointed snout vertically downwards, is required for this deception. The tiny fins serve only to maintain and adjust the razorfishes' position as they are moved to and fro with the weeds in the swell.

A more sinister disguise is that of the stone-fish, *Synanceja*, which lives in shallow water on coral reefs. The stonefish is an ugly, big-headed creature that spends most of its time lying on the bottom. Its shape and dark, warty skin gives it exactly the appearance of an en-crusted piece of dead coral. This disguise, combined with sluggish habits and a fondness for shallow water, makes the stonefish one of the most dangerous animals on the reef, for hidden in the dorsal fin is a sharp spine through which an extremely toxic venom can be injected.

The examples of disguise given so far depend on a much altered body-shape. But there is another method for achieving the same result —

Above: The Dragon fish is usually well disguised among coral growth, but if disturbed it can perform a dramatic threat display associated with a powerful sting in the dorsal fin.

Left: Another fish well known for its painful sting is the stone-fish, so called because it resembles a lump of old coral.

adventitious disguise. Many species deceive their enemies by constructing a canopy over themselves, or by plastering themselves with various materials. A few even plant live organisms on their own backs. It is a cumbersome form of protection, suited only to slow-moving or sessile (permanently-fixed) animals, but it can be extremely effective.

Adventitious disguise is often procured by a covering of materials gathered from the surfaces over which the animal moves. Case-building (the construction of mobile cases) has been evolved many times. In freshwater, caddis-fly nymphs make a variety of cases, depending on species. Some put together cases out of chewed-off pieces of reed. The huge family of bagworm moths constitute a parallel development on land; the larvae of some species make cases of *Acacia* spines, bark or lichen fragments, while others make use of 'frass', a mixture of wood-

dust and droppings left by wood-boring beetles. Cases such as these confer physical protection as well as disguise.

A simpler form of adventitious disguise involves sticking various materials on to the upper surface of the body. This allows greater freedom of movement than does a case, but gives far less physical protection. On rocky shores various crabs and prawns encrust themselves with pieces of algae or sponge. The pieces continue to grow and eventually provide a perfectly natural-looking disguise.

A more macabre example is provided by African assassin-bugs. The nymph of *Africanthus* fastens its victims' corpses to its back after sucking them dry. Its head and legs are plastered with grains of sand, but its eyes, antennae and the vicious 'beak' are kept scrupuously free of debris. The mass of shrivelled skins doesn't make the nymph look like anything in particu-

Below: Shoal of razor-fishes drifting head downwards like fronds of seaweed.

Right: The nymph of the African assassin-bug *Africanthus* is disguised as a lump of detritus. It covers itself with grains of sand and the corpses of insects it has sucked dry; only the eyes and sensitive antennae remain uncovered.

lar, but then, of what interest is a pile of empty insect skins to a hungry insectivorous bird? One might describe this as rubbish- (or trash-) disguise.

Another larva that covers itself with dead skins is that of the tortoise beetle, *Aspidomorpha*. But in this case the skins are its own. When they emerge from the egg-case the larvae feed rapidly and soon black droppings begin to accumulate in the tail of long hairs; when the larva moults, the shed skin (complete with excreta) sticks to the next instar's tail, and so on with every moult. By the time the larva is fully grown it carries over its back a mass of its former skins and droppings. Seen from above it certainly does not look good to eat. The pupa forms under the same cloak of accumulated remnants. Paradoxically, the adult beetle which eventually emerges is one of the most beautiful of all insects, being a brilliant iridescent gold all over.

Certain spiders display a form of adventitious disguise which might equally well be termed adventitious camouflage. Somewhere on the web a platform of thick silk is spun, or a piece of bark or leaf is incorporated into the structure. The spider then rests on the prepared background which it resembles in coloration. Instead of seeing a spider an observer notices only an empty web, or a fragment of leaf caught in one. Another ploy is to leave the sucked-dry victims gathered together in one part of the web and sit close to them looking like one of the meal packages. One species of *Argiope* in Africa makes a conspicuous white cross in its web with a special kind of flocculent silk. The web is built in thick vegetation near the ground so the main network of fine threads

is invisible. The white cross stands out, but is seemingly empty. On the other side of the web (facing the ground or close to other objects) sits the pale spider, its legs spread out two together along each arm of the cross. Many spiders employ more conventional adventitious disguise to protect their eggs. The egg-case is fastened to something solid and covered over with particles of bark, lichen or other material, held together with fine silk. The European *Cyclosa conica* makes a flat case attached to a twig and disguised with pieces of bark. After laying, the female sits close to the eggs till they hatch several weeks later. While she is protecting the eggs from ants and other marauders she is herself protected from birds by her own disguise as a broken twig-base.

The behaviour characteristic of disguised animals is similar to that of camouflaged species. They tend to be nocturnally active, or move about warily during the day ready to freeze or change to some other defence tactic if detected.

One peculiarity of many animals that rely on concealment is the extraordinary care with which they dispose of their excrement. No matter how well an animal is concealed by its colour or shape, an accumulation of faeces nearby could easily give it away. An individual resting high in a tree can simply allow its droppings to fall through the foliage, but one resting on the ground must take active steps to avoid the problem. An East African grasshopper, *Lobosceliana femoralis*, which rests on open ground disguised as a shrivelled brown leaf, has a complex ritual for getting rid of its droppings: whenever one is about to emerge, the grasshopper turns the tip of its abdomen upwards so that the elongate pellet emerges vertically, like a rocket coming up from an underground silo. Then, just as it seems about to topple, the insect leans slightly to one side, brings one of its hind legs forward and gives a mighty kick at the pellet, sending it several feet away. Seeing this normally sluggish insect make such a swift, precise movement comes as a shock and emphasizes the importance of this adaptation which must have taken long to perfect.

Baby forest antelopes such as the African bushbuck have an even greater problem. For long periods they lie still in dense vegetation protected from mammalian predators by their camouflage. From time to time the mother visits her baby and while it suckles she licks under its tail and eats the droppings it produces. The youngster's instinct to retain all faeces until the mother is present is very strong. Were they to be voided their smell would quickly attract a hyaena or other killer. In Africa, baby antelopes are frequently found lying alone in the forest and, believing them to be lost or aban-

doned, people misguidedly take them home. The young animal readily takes to the bottle, but often dies after a few days. The cause of death is constipation, for without the stimulus of the mother's licking the poor mite seems unable to empty its bowels. Incidentally, brushing its rear-end periodically with a moist paint-brush is all that is required to save its life.

Nest sanitation, as practised by many birds, is also highly ritualized. After it has been fed, the nestling produces a neat packet of white faecal matter which the parent immediately removes and drops far away. Apart from hygiene advantage, camouflaged or disguised young thereby avoid becoming surrounded by conspicuous and smelly droppings that might attract a weasel, stoat or fox. Sea-birds nesting on cliffs, woodland birds like the goldfinch and crossbill that nest in dense foliage high in trees, and birds like swifts and kingfishers nesting in crevices or tunnels do not practise nest-sanitation to the same degree; their nests become absolutely filthy, but this matters little since concealment is unimportant.

Below: Young harnessed antelopes, or bushbuck, rely on their camouflage to escape detection when left alone for long periods between meals. The mother keeps her offspring scrupulously clean to avoid any revealing odours.

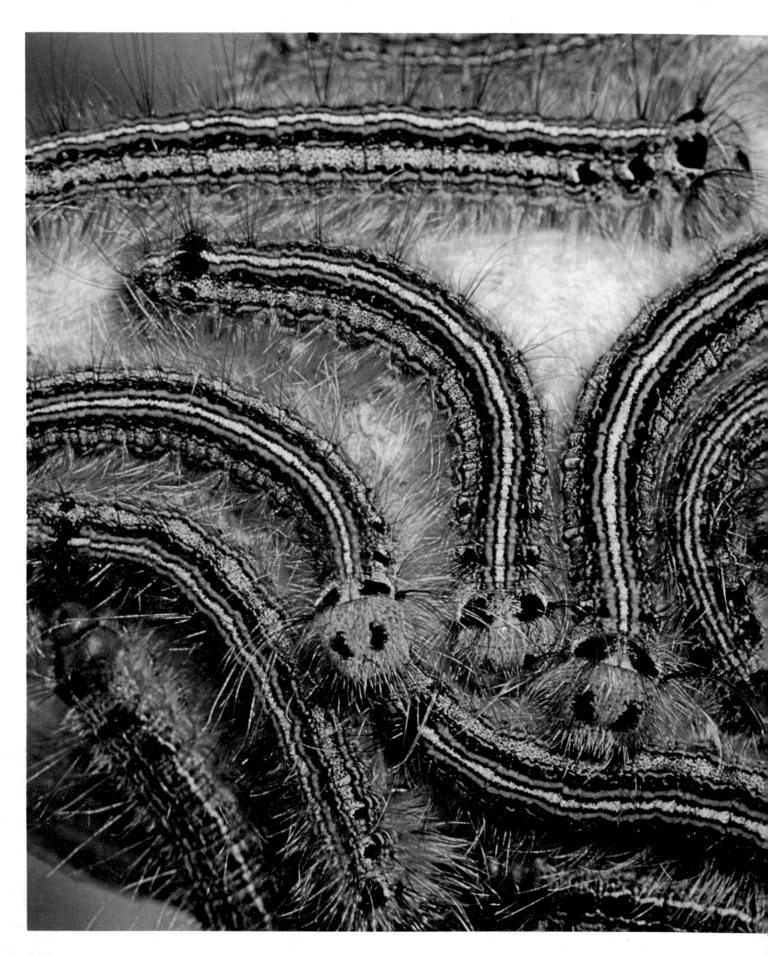

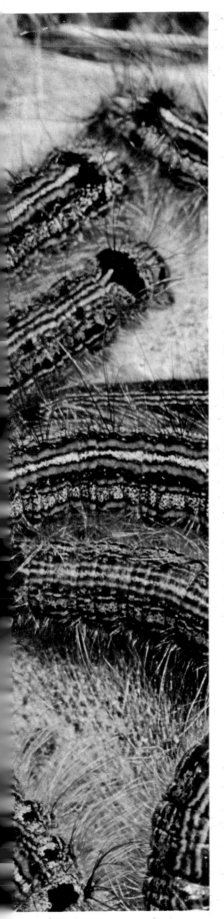

# Warning Colours

Warning colours serve to advertise an animal in a mainly green and brown world. Red, orange and yellow stand out best in most situations, while blue, black and white can be conspicuous in some surroundings. Warning colours are usually brilliant or, as an artist might describe them, 'saturated'. They are produced either by pigments such as carotenoids, or as with the metallic purples, greens and gold worn by many unpalatable beetles, by physical effects. Iridescent warning colours are most effective in strong illumination and tend to be adopted by species that thrive in full sun.

Some aposematic animals are uniformly coloured, but most have a bold pattern incorporating two or more colours. This makes them hard to miss anywhere. Because certain combinations and patterns are particularly effective they have been adopted by species not closely related to one another. For instance, black and yellow stripes characterize many wasps, arrow-poison frogs, cinnebar moth caterpillars and some distasteful grasshoppers. Bold spots on a plain contrasting background, as exemplified by ladybird beetles, is another inherently conspicuous pattern. Ladybirds illustrate the point that, just as unrelated animals may arrive at the same warning pattern by 'convergent evolution', closely related species often develop different patterns. In Europe, the 22-spot ladybird, *Psyllobora 22-punctata*, has

Caterpillars of the European lackey moth assembled on the silk tent they build communally among the foliage. Grouping enhances their conspicuousness and the white background renders their warning colours more prominent, especially when they become dull during skin-changes.

black spots on a yellow ground, while two-spot ladybirds, *Adalia bipunctata*, have black spots on a red ground, or, as a common variant, red spots on black.

Red, the colour that symbolizes danger so well on land, is quite ineffective in water. The longer wavelengths of light do not penetrate far below the surface. Most of the red is filtered out in the upper 10–20 metres, followed soon by orange and yellow. Only the shorter green and blue wavelengths reach to any great depth, which is why in deep water all pale-toned animals look blue. This applies even to species which are seen to be white or yellow when brought to the surface. Deep-water organisms with a great deal of red in the skin are effectively almost as black as if coloured by a black pigment. There is no red light to be reflected and red pigments absorb blue and green light. Interpretation of marine animals' coloration is made even more difficult by the varying adaptation of the predators' eyes at different depths. Surface-living and shallow-water fishes have good colour vision, but those which live deeper down where colour appreciation is valueless do not. They have given it up in favour of more sensitive monochromatic vision. The situation parallels that on land where animals active by day see colours, while their nocturnal relatives have eyes rich in light-sensitive cells but devoid of colour receptors.

Because of these difficulties it is safer to concentrate on patterns rather than colours when considering the defensive adaptations of any other than surface-living species. The spectacular red-and-white sea-dragon, or lion-fish, *Pterois volitans*, looks as if it is warningly-coloured when seen in an aquarium. Moreover, this seems a plausible interpretation for, like other scorpion-fishes, the sea-dragon has an

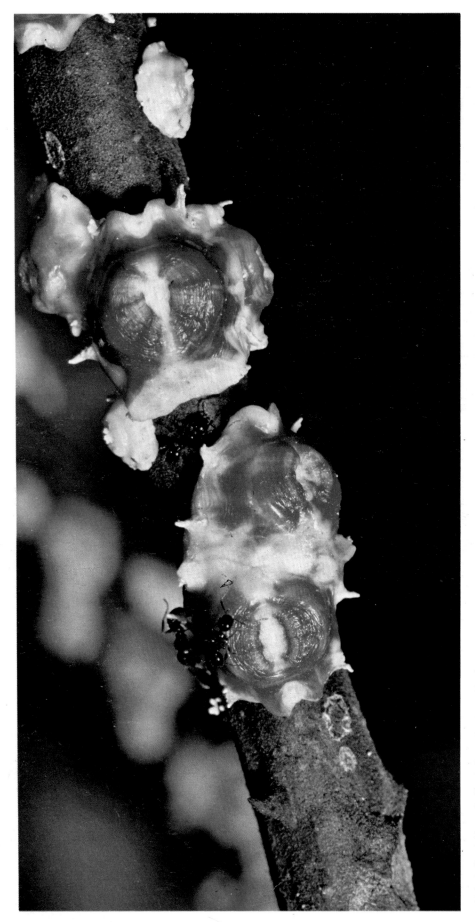

extremely venomous sting in its dorsal fin. But if we forget the red colour and imagine just dark stripes, as we would see them in the sea, the striping is obviously disruptive. There is even an eye-stripe to merge the pupil into the overall pattern.

There are, however, many fishes that carry unambiguous warning colours: puffer-fishes with inflatable prickly skins and poisonous flesh, venomous weevers, *Trachinus*, and clown-fishes, *Amphiprion*. The latter take their name from their vivid orange, black and white patch-work coloration which makes them extremely conspicuous in the well-lit coral gardens they inhabit. They are probably advertising their well-defended surroundings; the clown-fishes spend their lives in a close, symbiotic relationship with large *Stoichactis* sea-anemones. They swim among the tentacles of the anemone with impunity, protected by a mucous that inhibits the stinging cells. If artificially separated from its anemone a clown-fish is soon attacked, showing it is not unpalatable to predators. What, if anything, the anemone gets out of the association is a debatable point.

With few exceptions like the *Amphiprion* fishes, which rely on the weapons of others, warning colours are worn by animals that have a potent defence. This may be chemical as with the many poisonous species, or physical in the form of armour or sharp weapons. There are also the mimics, to be dealt with later, which

Far left: The distasteful wax coating of *Ceroplastes* scale-insects is advertised by their bright pink colour. The attendant ants may offer further protection in return for the honey-dew secretion they collect.

Left: Slug-caterpillar covered with stinging hairs that can inflict painful wounds, West Africa.

Below: The noxious skin secretion of the arrow poison frog *Dendrobates leucomelas* acts as a repellent. When concentrated by South American indians the toxin becomes a deadly venom.

Below right: A flannel-moth caterpillar from Central America. The beautiful golden hairs carry an extremely powerful irritant.

wear the same colour but are, in fact, quite harmless.

Chemical defence depends on toxins, produced in most cases by special poison glands. 'Venoms' are one class of toxin. They are injected by a sting or poison fang into the skin of an opponent and are usually the most potent. 'Crinotoxins' are secretions on the surface of an animal; they may be corrosive or obnoxious but are rarely lethal. The chemistry of toxins is extremely complicated. Relatively few have been analysed, but they seem mostly to be large, complicated molecules; many are proteins. Unless they are used also for killing prey, as with snakes, scorpions, spiders and jelly-fishes, poisons used for defence do not usually cause permanent injury. It is sufficient that a predator be given pain, or a disagreeable taste in its mouth, for it to give up the attack and let the victim go. The predator then associates the alarming experience with the distinctive colours and patterns of the victim – and studiously avoids that species in future.

Often we receive the wrong impression of the potency of toxins from the way man employs them. For instance, the skin secretion of arrow-poison frogs in the genus *Dendrobates* has long been used by South American hunters as a lethal poison. The tiny frogs are caught in large numbers and slowly roasted alive over a fire. The thick liquid which oozes from the skin is collected and concentrated by evaporation before being applied to the tips of arrows. A small mammal or bird hit by a poisoned arrow is immediately paralysed. Man is using the frog's crinotoxin as a deadly venom, whereas its natural function is as a distasteful substance and vomitiferant. In other words, it merely drives a predator to spit the frog out of its mouth without causing it any lasting harm.

Most of the venomous insects belong to one class, the Hymenoptera, which includes ants, bees and wasps. There are also stinging caterpillars, like the sabines of tropical America, which have a battery of extremely unpleasant stings all over their bodies. Honey-bee venom, used only for defence, is a veritable witches' brew. It contains proteins that affect the nervous system and respiration, as well as causing local swelling round the sting and burning sensations. Other stinging Hymenoptera have similar venoms, but ants produce a different mixture rich in formic acid. The nature of caterpillar venoms is largely unknown though some, like that of the American puss caterpillar, *Megalopyge*, cause humans intense pain that lasts for

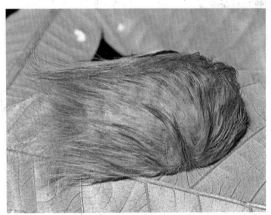

days. The venom is contained in their attractive long golden 'fur', fine hairs which can penetrate the skin.

All spiders inflict a poison when they bite their prey, and a few have a venom so potent it can be used in defence against large predators. Best known is the American black widow, *Latrodectus mactans*, but other members of the genus are equally dangerous to man. In Australia the 'night-stinger', *L. hasseltii*, is much feared and has led to many fatalities, mostly among children. Africa has the button-spider, *L. indistinctus*, and around the Mediterranean lives another race of the black widow. The *Latrodectus* spiders are all small-bodied, timid animals. Their aim is to be left alone, hence their black and red colours. Other spiders are either known to be dangerous or suspected

Right: Thorny-bodied spiders, like this Mexican *Gasteracantha*, show off their hard spines with brilliant colours.

of being so. It is therefore prudent to treat any species bearing warning colours, however small, with respect. The large 'tarantula' spiders are mostly docile, but some are exceedingly dangerous due to the large quantity of venom they inject.

The eight hundred species of scorpions are all venomous. The poison is used sparingly when feeding and if a victim can be torn apart easily it is not stung at all, but if it is large and struggles it is pacified with a quick jab. Scorpions also use their stings for protection. Some species, like the African fat-tailed scorpion, *Androctonus*, can inflict in man an excruciatingly painful sting which often proves fatal. Scorpions are not obviously aposematic, though some species are uniformly black or pale-yellow.

A wider variety of stinging invertebrates occur in the sea. Thousands of corals, anemones and jelly-fishes, coloured brightly to warn, are armed with nematocysts (stinging cells). These serve partly to stun small prey, partly for defence. An Australian jelly-fish, the sea-wasp, *Chironex fleckeri*, produces an extremely virulent heart poison. It is reckoned to be the most venomous of all creatures. There are also extremely poisonous octopuses and sea-snails. In the ocean depths are venomous invertebrates that give warning by emitting light.

Many fishes are able to inject poison through modified fin-rays, as exemplified by the stone-fish referred to earlier. Alternatively the poison spine can be on the gill cover (weever-fish), or on the tail (sting-ray). Another weapon is high-voltage electricity used against predators by electric catfish and torpedo rays. While many of these dangerous fishes are strongly patterned, some are cryptic and use their weapon as a second-line defence.

Among reptiles, only two venomous lizards are known: the pink-and-black gila monster, *Heloderma suspectum*, and the bearded lizard, *H. horridum*, which has black and yellow stripes. Both live in the deserts of Mexico and neighbouring U.S. states. As already mentioned, most snakes, including the venomous kinds, are cryptic. Among the few blatantly aposematic species are the yellow-bellied sea-snake and the coral snakes. Some of the cryptic snakes perform a warning display when threatened and cobras show strong patterns on the hood when they display. Snake venoms vary greatly in toxicity; some induce respiratory failure while others destroy the blood.

Shrews are mildly venomous mammals, some being black or black-and-white, possibly for warning. Their poison glands are associated with the lower incisors. The male duck-billed platypus of Australia injects poison with a spur on its hind leg. Though it might seem surprising, hedgehogs may also now be regarded as venomous mammals in view of recent discoveries concerning their defensive behaviour. These will be described later.

Crinotoxins are far more widespread than venoms, but little is known about their mode of action. This is because the major research effort by toxicologists has been devoted to venoms which are of more consequence to man. Most crinotoxins are liquid, though in some cases the liquid volatilizes in the air to produce a defensive gas. In at least one example the gas actually explodes. The secretions generally arise from glands in the skin which connect through pores to the surface. Alternatively, the toxins are manufactured in glands connecting with the mouth or anus, and emerge from these openings when required for defence. A few animals have poisonous blood and secrete by voluntary bleeding.

Among terrestrial invertebrates, a wide variety of species utilize defensive secretions. Even some harmless looking wood-lice (or sow-bugs) can exude a mild deterrent, said to smell like butyric acid. Millepedes produce more potent chemicals and have accordingly become warningly-coloured. Most are a uniform black, yellow or orange, though there are tropical species gaily patterned in black and pink. Some millipede secretions are highly corrosive to the skin of other animals, while others appear to depend on an unpleasant smell and taste.

Right: Bright blue spots give warning of the torpedo ray's ability to produce a powerful electric shock.

Below: Tropical millepedes, often brightly-coloured, produce blistering skin secretions containing acids and cyanide.

Overleaf: Pacific black and yellow sea-snake. The warning patterns of sea-snakes are probably recognized instinctively by potential enemies.

Iodine, chlorine and hydrogen cyanide have been detected in the exudates of various species, as well as hydroquinone, an extremely bitter substance. At least one millipede can discharge its poison a distance of 20 in., but most simply let it ooze on to the body surface when they are molested.

Spiders do not rely on crinotoxins, though some are probably distasteful. A few spiders use their venomous bite for defence. Other arachnids do have defensive secretions, however: the whip-scorpion, *Mastigoproctus giganteus*, sprays a liquid containing 84 per cent acetic acid. A further 5 per cent is a 'spreader' which makes the strong acid flow over the integument of an attacker and aids its penetration. How the whip-scorpion can store such a corrosive liquid in its body is something of a mystery. Some harvestmen and mites are also known to employ chemicals for defence and many are brilliantly coloured.

There are countless thousands of insect species that produce some kind of crinotoxin. A large proportion of them are aposematic. Blister-beetles produce a substance called cantharadin which produces severe and persistant blistering when the insect is pressed against the skin. A bird or other predator grabbing one of these beetles soon drops it, showing obvious signs of distress. The cantharadin used as a drug by veterinarians is

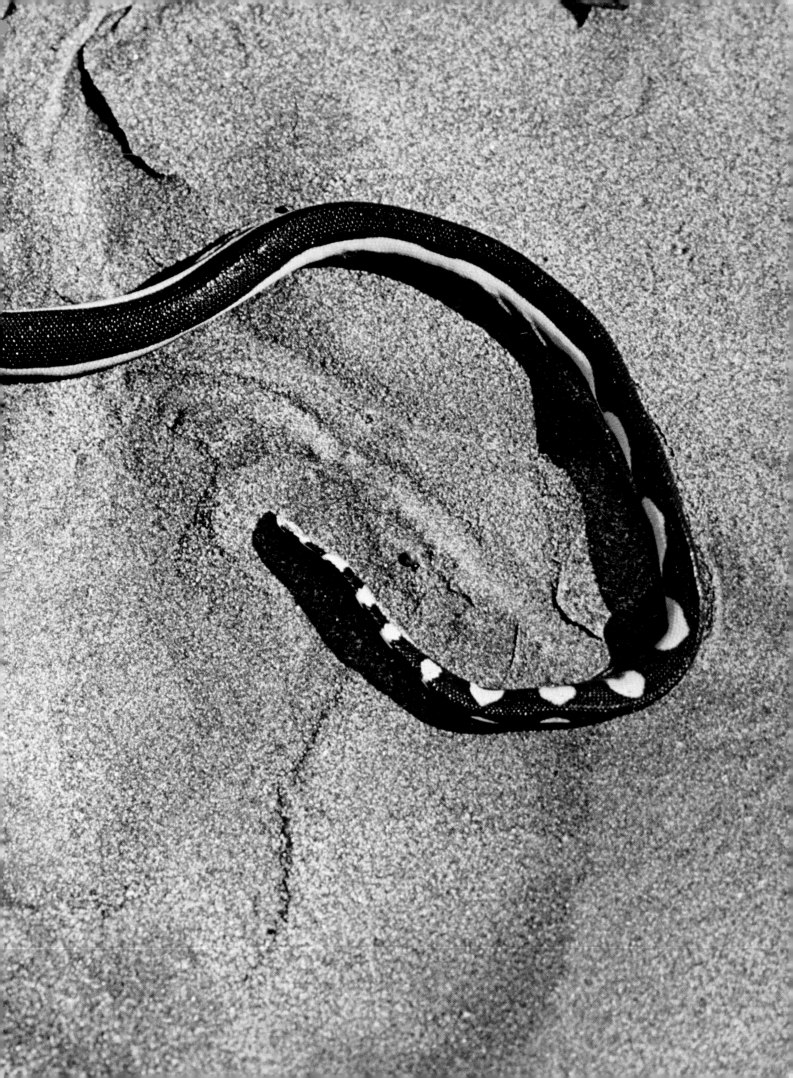

extracted from a member of the blister-beetle family, the Spanish fly, *Lytta vesicatoria*. Some beetles are known to squirt formic acid and other irritant or foetid-smelling liquids in the face of an attacker. But most remarkable are the red-and-black bombardier beetles, *Brachinus*. Their defensive chemistry has only recently been investigated, though they have long been known for the popping sounds from which they obtain their name.

When it is under attack the bombardier beetle turns away and ejects a hot blast of gas into the face of its advesary. Two sets of glands secrete different chemicals — hydrogen peroxide and hydroquinone — into a reaction chamber. The reaction generates a great deal of heat and the gas reaches a blistering 100°C as it explodes into the air. Small rodents and birds as well as ants are known to react adversely to the discharge.

Grasshoppers in the family Pyrgomorphidae are also able to project liquids, foul-smelling and bitter-tasting, into an enemy's face. Others in the same family, like the red-and-black *Dictyophorus griseus* of East Africa, produce a defensive foam. When molested, the grasshopper secretes a yellow liquid from glands between the segments. As it flows across the external openings of the respiratory system the liquid is turned into a thick, yellow froth by the air forced through it. Apart from the yellow colour added to the red-and-black warning pattern, and the acrid smell, there is a shock element to the display: the foam appears very suddenly with a squirting noise audible even to humans several feet away.

Defensive secretions are relied upon also by cold-blooded vertebrates. The sea lamprey, *Petromyzon marinus*, and the well-named soap-fish, *Rypticus saponaceus*, are coated with a noxious slime. Predators suffer considerable irritation of the oral membranes if they make the mistake of trying to eat one. The vivid black-and-yellow fire salamander, *Salamandra salamandra*, of southern Europe, like many other salamanders, also have noxious properties.

The most poisonous frogs are brightly coloured, and none more so than the arrow-poison frogs, *Dendrobates*, of Central and South America. The many species exhibit a wide range of warning patterns: black-and-yellow stripes; black spots on a red or yellow ground; bold patches of sky-blue and black; and various other combinations. In the Old World tropics, *Hyperolius* tree-frogs occupy a similar niche, though they are not so brightly coloured. Toads in the genus *Bufo* produce protective skin secretions, but we must assume they are not very potent for most toads are cryptic and although they are shunned by some predators others will eat them with impunity. One African snake, the night adder *Causus rhom-*

*beatus*, even specializes in toads. In fact it has become a common town snake in areas where *Bufo regularis* has taken to hunting for insects near houses and street-lights. Incidentally, the much larger file snake, *Mehelya capensis*, is known to eat both the toads and the night adders.

Among mammals, noxious secretions are produced by members of the family Mustelidae (badgers, polecats, weasels and allies). The fluid is produced in the anal glands and derived from the scent used in minute quantities for marking territorial boundaries. The skunks of North and South America are notorious for their ability to eject a persistently evil-smelling liquid when molested or threatened. They have a parallel in Africa in the zorilla, or striped weasel, *Ictonyx striatus*, which defends itself in the same fashion. Like skunks, zorillas have a bold black-and-white pattern and move about noisily at night. The anal secretions apparently afford wide-ranging defence against mammalian predators and have the special attribute of persistence, often for days. Several large members of the same family also advertise themselves prominently. The eight species of badgers, including the European badger, *Meles meles*, and American badger, *Taxidea taxus*, are all powerfully-built animals with black and white stripes on the head. Like the smaller mustelines, badgers have anal glands and the well-named stink-badgers, *Mydaus*, of tropical Asia rely on their secretion for defence. The ratel, *Mellivora capensis*, sometimes called the

Above: When disturbed the East African grasshopper *Dictyophorus griseus* produces a foul-smelling foam to emphasize its warning colours. These sluggish insects contain several powerful poisons, some derived from their food-plants.

African honey-badger, is white on the top of its head and back, and black underneath. Ratels feed mainly at night, but are sometimes active by day, especially when following the honey-guide, *Indicator indicator*. This is the bird that by repeated calls and short, guiding flights leads ratels (or humans) to a bees' nests it has found. When the nest has been broken into and the ratel has taken its fill of grubs and honey, the bird eats the wax and anything else that remains.

The larger badgers and the ratel (which is not a true badger) advertise not so much their smell as their other weapons. For although normally inoffensive creatures they can defend themselves vigorously if attacked. With powerful claws, sharp teeth and a tendency to hold on and shake when biting, they can inflict terrible wounds on animals far larger than themselves. Apart from great courage they are remarkable for their tenacity to cling to life, being able to withstand blows to an amazing degree and continue fighting even when severely wounded. What they are indicating to potential enemies, such as wolves or big cats, is 'I may look slow and easy to catch, but you'll certainly get hurt if you try.'

These are just a few of the chemical defence systems evolved by animals. Clearly there is great variety in the toxins used and it is interesting to look at where they come from. Most toxins are synthesized in the animals' own bodies or obtained second-hand from plants; few instances are known where the source is another species of animal.

Many plants manufacture toxic compounds and store them in their leaves as protection from herbivores. Old leaves tend to contain more than younger leaves and this is one reason why insects favour young growth. Unripe fruits may also contain toxins to prevent the plant's dispersal agents from taking them too soon; the toxins are transported out of the fruit when they ripen. The chemicals give a good measure of protection but, as always, some animals have broken through the defences. Even such poisonous plants as tobacco, foxglove, hemlock and laurel have their insect pests.

Having evolved a means of coping with its food-plant's toxins, an animal can begin to make use of the chemicals. By storing them in their bodies, they render themselves inedible to birds and other predators. The last step in

Below: Black and white patterns are commonly used for advertisement by mammals. The striped skunk produces a copious evil-smelling secretion from its anal glands.

this evolutionary sequence is the development of warning coloration. Presumably this can only be taken when the animal has become so obnoxious to most predators that it gains more from exhibiting itself than it loses by being easy to find.

The value of toxins, whether 'borrowed' from plants or newly manufactured, is so great that they are often passed on from parent to offspring. Caterpillars of the large white butterfly, *Pieris brassicae*, concentrate sinigrin and mustard oils obtained from members of the cabbage family. They can sequester enough to protect themselves not only as larvae but also throughout the subsequent pupal and adult stages. Thus the butterfly is protected by

mustard oils it accumulated weeks earlier when it was a caterpillar, and some is incorporated in the eggs the butterfly goes on to lay. In a similar way the American monarch butterfly, *Danaus plexippus*, obtains its cardenolide heart poisons from larval food-plants in the milkweed family.

One of the most intriguing examples of an animal that gets its poison from another is provided by the sea-slugs, or nudibranchs. These strange, soft-bodied molluscs rank among the most beautiful creatures in the sea. Their vivid warning coloration and lack of protective shell is explained by batteries of stinging cells able to deter most enemies. But these cells are not the sea-slugs' own; they once formed part of the bodies of sea anemones, corals or jelly-fishes eaten by the sea-slugs. In some mysterious way a sea-slug can eat its prey without the defensive stinging cells firing off – as they would if any other predator were involved. Even more amazing, the stinging cells are somehow sorted from the rest of the victim's body, which is digested, and passed through the gut wall. They must then migrate through the sea-slug's tissues until their new location is reached. All nudibranchs are not protected in this way; some equally colourful species defend them-sevles by secreting sulphuric acid.

The nudibranch story has been known for many years, but an equally fascinating one, involving two quite commonplace animals, was reported only recently. It appears that several hedgehogs, including the African

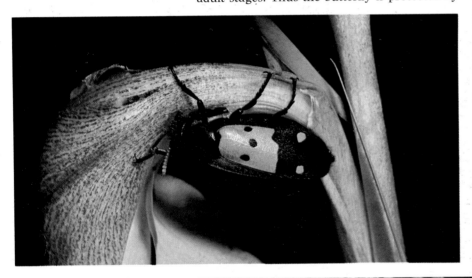

Above: Noxious insects may derive both poisons and warning colour pigments from their food plants. This African beetle is feeding on a lily rich in toxins and yellow carotenes.

Right: Brightly coloured sea-slugs, or Nudibranchs, like this *Flabelligera iodinea* from California, are armed with stinging cells derived from the tentacles of sea-anemones they have eaten.

Right: The African hedgehog has a black and white warning pattern and moves about noisily to make itself conspicuous at night. The sharp spines are annointed with irritant poisons derived from toads.

*Erinaceus pruneri,* anoint their spines with an irritant fluid derived from their prey. On catching a toad a hedgehog first bites off the head where the large parotid glands are situated. Much of the bufotoxin used by the toad in its own defence is stored in these glands. The poison does not deter the hedgehog, but does stimulate copious salivation. It then proceeds to lick the mixture of saliva and 'toad-juice' on to its spines. The biologist who discovered this remarkable behaviour performed experiments with human volunteers. While clean spines jabbed into their skin produced only a short-lived pain, the anointed spines gave a burning sensation that lasted for hours. Hedgehogs advertise themselves by moving about noisily and some species have a conspicuous black and white pattern.

As far as is known, the porcupine's quills offer only mechanical defence, but this is so effective that the quills are banded black-and-white as warning coloration. They are only loosely attached to their owner and once embedded in an attacker's skin they work their way in deeply, leading to painful festering wounds. Other animals with piercing weapons that rely primarily on crypsis reserve their weapons for emergencies, have the colour associated with the weapon itself. In this way it is noticeable only at close quarters. Antelopes rely first and foremost on camouflage to escape their many enemies and if this fails they generally take to flight. Only if they are cornered will they stand and fight, and then the sharp horns show prominently. Light-skinned antelopes such as reedbuck, impala, hartebeeste and gazelles tend to have blackish horns. Moreover the tip is often pale, as if accentuating its sharpness. On the other hand dark antelopes such as wildebeeste, bison and musk-ox tend to have pale horns with a dark tip.

Where the weapon can be hidden when not in use it can be conspicuously coloured without an animal's camouflage being spoiled. Many mammals that use their biting powers for defence have inherently conspicuous white teeth set against strongly contrasting gums and tongue. Usually these are red, but sometimes black or blue. Interestingly, the incisors of many rodents are bright yellow-orange in colour. This may be incidental, but it could signify a warning.

Despite an overall cryptic garb, many falcons, hawks and eagles have brightly coloured feet

Left: *Hyperolius pusillus* is a palatable tree-frog, relying on camouflage for protection. When disturbed, it leaps into thick vegetation and again 'freezes'.

and bills. These offer their main means of defence in close encounters. Most commonly, black talons contrast with yellow or orange-red feet, while the beak is yellow or red, tipped with black.

Invertebrates also may have hidden weapons. In tropical Africa there are large caterpillars in the family Lymantriidae that rest on tree trunks protected by their camouflage. They have fine hairs along their sides to help the body merge with the bark. Normally they lie still all day, but if one of them is disturbed it suddenly displays transverse tufts of crimson hairs previously hidden in skin-folds. At the same time it vibrates its body to arouse its neighbours, who also erect their weapons. If a bird or other predator grabs at the caterpillar's head, it risks getting a mouthful of the irritant hairs. Should it aim at the rear, the caterpillar simply throws its head over backwards, thrusting the hairs into its adversary's face.

The behaviour characteristic of animals which display permanent warning colours is very different from that of cryptic species. They tend to be active by day, or if they must feed at night they rest through the daylight hours in prominent places. When disturbed an aposematic animal often begins to move slowly, instead of 'freezing' or taking to flight. The difference in behaviour is exemplified by two closely related tree-frogs inhabiting the same East African forest. One is a small green frog, *Hyperolius pusillus*, which is known to be palatable to birds. It rests by day under a leaf with its body pressed tightly against the surface and legs tucked well in, blending perfectly with the background. Left alone, it remains quite

still all day, but if touched the reaction is immediate. From its crouched position the tiny frog suddenly leaps a metre or more into the surrounding dense vegetation where it again crouches in the cryptic position. The other species, *Hyperolius marmoratus*, is a near relative, but very different in appearance: its head and back are spotted with black and white, while the belly and legs are bright red. This is, of course, typical warning coloration and is associated with known toxic properties. During the day *H. marmoratus* sits on top of a flat leaf in an alert position with its long toes splayed out. When approached closely it does not leap to safety but crawls about and climbs laboriously, displaying its pattern prominently. The marbled tree-frog can be pushed into making a leap, but instead of launching itself far into the bushes it aims at a nearby twig where it hangs precariously like a trapeze artist. This allows the red underside to show to best advantage. A bewildering variety of amphibians live in the rainforests of tropical America, most of them nocturnal, but the garishly-patterned arrow-poison frogs are to be seen clambering about in the leaf-litter throughout the day, seemingly oblivious of all danger.

Another common trait accompanying warning colours is an apparent lack of sensitivity, to the extent that the animals appear clumsy. They tend to get on with whatever they're doing without maintaining a constant lookout. The aposematic hedgehogs, badgers and skunks all plunge through the undergrowth sniffing and snorting, rarely looking up during the quest for food. True, unlike most aposematic species they are nocturnal, but this is because their prey are readily available only at night. The actions of such animals are quite different from those of deer, rabbits and other creatures with many enemies. These are ever-vigilant, pausing frequently to look and listen as they move silently through the vegetation. At all times they are poised ready to take flight at the slightest sign of danger.

Conspicuousness is also a characteristic of small, warningly-coloured animals. Woolly-bears, the hairy caterpillars of tiger moths, can often be seen moving about during the day. The same is true of slugs and millepedes, provided the weather is wet enough for them. One of the most dramatic examples of this single-mindedness is provided by the bird-wing butterflies, *Ornithoptera*, of South-east Asia. These huge insects sail lugubriously across the forest clearings displaying their gaily patterned wings in slow flight. This may look like an open invitation to insectivorous birds, but it is, in fact, a warning that they are highly distasteful. The equally poisonous and brilliantly coloured *Heliconius* butterflies of the American

tropics have the same inviting, slow, sailing flight.

Since the aim of aposematic species is to be seen it isn't surprising to learn that they often gather together. The most familiar assemblies of this kind are formed by ladybird beetles as they prepare to hibernate. Another common sight in northern Europe are masses of black peacock butterfly caterpillars in nettles, their food plant. Evidently peacock caterpillars are unpalatable to most birds, though as so often happens their defence has been beaten by at least one predator – the cuckoo eats these and other 'protected' caterpillars with impunity. Other caterpillars, grasshoppers, saw-fly larvae and stinkbugs with warning colours also mass together. Often the gregarious behaviour is limited to the young stages, but in certain butterflies the habit persists throughout adult life.

The biggest cluster of all is formed by monarch butterflies, *Danaus plexippus*, in their wintering quarters. Countless millions of them assemble from all over North America in a small mountainous region in Mexico where they literally smother the trees in a vivid orange blanket. No other butterfly assembly of this magnitude is known, but there are many distasteful butterflies which form small, communal roosts at night. The habit is common to milkweed and *Heliconius* butterflies in the Americas and to species of *Acraea* in tropical Africa. Before settling down together for the night they may perform an aerial dance around the chosen spot. Doubtlessly this conspicuous display serves to attract other individuals towards the roost, making it as large – and therefore as well defended by odour – as possible.

The examples cited above are all insects, but there are also millipedes, like the African *Habrodesmus* that swarm for better protection. Some fishes also unite into tight, highly conspicuous shoals to emphasize warning colours associated with noxious qualities.

The advantages of grouping together stem from enhanced visual conspicuousness and, perhaps more importantly, a concentration of the characteristic warning smell. For some species the advantages are evidently greater for the young than they are for older individuals. The first instars of distasteful *Dictyophorus* grasshoppers remain clustered in a black mass on their food plant. They are distasteful, but they are also very small. Despite its black coloration, a lone individual could easily be eaten by mistake, and though it would probably be spat out again it could be seriously harmed. Massed together, the little grasshoppers are unmistakable. Their acrid smell cannot be detected by birds but it is readily detectable by mammalian predators hunting at night. The nymphs disperse when they become large enough to display their warning colours

Below: The closely related *Hyperolius marmoratus* is poisonous and rests in conspicuous situations. If disturbed it clambers about laboriously showing its red legs and belly to best advantage.

Right: Rapid colour changes are employed by venomous species of octopus as a dramatic form of warning display.

prominently and frighten enemies by a striking threat-display.

One question that readily comes to mind as we look in detail at warning coloration is 'How do predators learn to associate particular colours, patterns and kinds of behaviour with unpleasant stimuli?' For many years after this question was first asked only anecdotal evidence supported the opinion of naturalists that the predators had to learn the association by bitter experience. Now there is experimental proof that predators do indeed have to learn. They learn rapidly early in life, and have a long memory as the following example shows.

In a typical experiment, a naïve toad, raised in captivity on a diet of palatable insects, is suddenly presented with a wasp. Invariably the toad strikes at the wasp and is immediately stung in the mouth. The poor toad reacts violently to this by spitting out the wasp and rubbing frantically at its mouth — as if to rub away the pain. It is then fed on palatable insects again for a while, and the experiment is repeated. Some individuals are such good learners that one experience with a wasp is enough. Even the poorest learners need only a few nasty shocks before steadfastly refusing to seize wasps — even when hungry. These experiments seem cruel until we realize that every wild toad must go through the same procedure as part of the 'growing-up process'.

Similar experiments have been carried out with predacious mammals, birds, reptiles and fishes offered a wide variety of potential food items. From the results we can generalize that predators do not have an instinctive fear of aposematic prey, but must learn to associate the colours, patterns, behaviour and smell of such animals with their unpleasant attributes.

Observations have also revealed marked variances between the reaction of different predators to the same noxious species: bees and wasps are avoided by most insectivorous birds which soon learn that black-and-yellow striped insects can sting, yet bee-eaters eat them readily. Bee-eaters have an instinctive ability to de-sting bees by knocking them repeatedly against a branch before swallowing them. The knocking completely destroys the stinging apparatus so that the venom, which is harmless when swallowed, cannot be injected. This behaviour is, however, exceptional; there is abundant evidence that stinging hymenoptera, ladybirds and other aposematic insects derive a great deal of protection against most vertebrate predators.

If we look at the experiments from the viewpoint of the prey another interesting fact emerges: warningly-coloured animals generally survive being seized, mouthed and regurgitated. They tend to have a thick integument, and their internal organs seem to be exceptionally tolerant of being crushed. Most lepidopterists have at some time had the distressing experience of seeing one of their specimens waving its antennae after several days on the setting-board. It has come round after spending the usual time exposed to lethal fumes in the killing-jar and being transfixed by a pin through the thorax. Such tenacity to life is an important attribute, for a warning pattern can only deter experienced predators.

Another question one might well pose is 'How did animals evolve warning coloration in the first place?' In some cases the answer is that presumably they are derived from a single unpalatable ancestor, which by adaptive radiation gave rise to the present variety of species; but a different explanation is required for situations where only a few members of a family are aposematic while the rest are cryptic and palatable. For example, of the many kinds of octopus in the world the majority are masters of camouflage and can change colour to suit the background in a matter of seconds; if this evasion fails it escapes in a cloud of ink. Yet one Pacific species, the blue-spotted octopus, *Octopus maculosus*, has quite a different defence system. Flashing its bright colours in an ever-changing threat-display, it warns all-comers that it is extremely venomous. Its jaws connect with glands that secrete a neuromuscular poison that is lethal to humans (though to marine predators it may be less toxic). Evidently in this case the warningly-coloured species has evolved from cryptic ancestors.

The anatomy of the blue-spotted octopus's poison glands gives a clue as to how the transformation occurred. They are nothing more than modified salivary glands. Presumably along one line of *Octopus* evolution there was an increasing specialization in the capture of active prey. To prevent their escape they had to be stunned with the first bite. Natural selection therefore favoured individuals whose saliva contained the most toxin. At first the octopuses were probably only mildly venomous and their main defence against enemies remained concealment. But gradually, through evolution, the toxin became more potent, and the poison apparatus more specialized. Eventually the stage was reached where the ancestral species was able to rely on its poison for defence. Crypsis was no longer the most effective coloration, since it often led to their being mistaken for non-poisonous octopuses. So natural selection began to favour those mutants with bright colours that were formerly selected out because they were conspicuous. Distinctive colours, permitting easy recognition, were selected for — and the present blue-spotted octopus arose as a new species.

A similar evolutionary path may be envisaged for aposematic moths and butterflies belonging to families most members of which are cryptic. For these the first step, made by a cryptic ancestor, was the breaking of some plant's poison-barrier against insect attack. Initially the only advantage to the caterpillars in being able to feed on a toxic plant lay in their having little competition for the food. Eventually, though, the plant-derived toxins in their bodies also began to confer advantages. Selection then favoured those individuals storing the most toxin and finally those which were also conspicuously coloured.

Evolutionary processes such as these take thousands or millions of years. They result not in perfection but in the best compromise. The caterpillars of the magpie moth, *Abraxas grossulariata*, and garden tiger, *Arctia caja*, have developed showy colours, the magpie to advertise its poisonous nature, the tiger its irritant hairs. We may presume that both derive a large measure of protection against insectivorous birds generally, yet both are eaten by one species, the cuckoo. In relation to cuckoos their showy appearance must be a considerable disadvantage, but the point is that cuckoos are not really common while the many bird species which would eat them if they were not aposematic are abundant.

# Mimicry

The restrictive definition of a mimic, as used here, is an animal that has clearly evolved in imitation of another species – called the model. Excluded by this definition are all cases where an inanimate object is imitated. Other authors may call them 'stick-mimics', etc., but here they fall in the category of disguise.

The simplest type of mimetic resemblance is where some palatable species has assumed the appearance (and often behaviour) of an unrelated species which is avoided by predators. The North American viceroy butterfly, *Limenitis archippus*, for instance, looks just like the monarch, *Danaus plexippus*. Not only are they patterned alike but they share the same slow, confident way of flying. Often one is mistakenly identified as the other. This pair of butterflies are not simply alike because they are closely related. The monarch is one of the 'tiger' butterflies (family Danaidae), which are all similar in appearance despite a world-wide distribution. The viceroy, on the other hand, belongs to the Nymphalidae and is quite unlike other members of the family (fritillaries, painted lady, white admiral, etc.).

There can be no doubt that the viceroy has evolved as a mimic of the monarch. North Americans often call it simply 'the mimic butterfly'. The value of the mimicry to the viceroy is not hard to appreciate, for the monarch is known to be highly distasteful. Stored in its body are toxic chemicals which it

Female Eurasia cuckoo in the act of laying in a reed warbler's nest after first removing one of the host's eggs which it holds in its beak.
Cuckoos mimic hawks in their general appearance and have also evolved egg-mimicry.

accumulates as a caterpillar. The food plant is milkweed *Asclepia*, which is rich in cardenolide heart poisons. Incidentally, the milkweed is presumed to manufacture the cardenolides for protection from herbivorous mammals and insects. The monarch caterpillar has developed tolerance of the toxins, and now puts them to good use for its own protection. Caterpillars of the viceroy feed on leaves of poplars and willows. They derive no toxins from their food and, unlike the monarch's larvae, must feed at night to avoid bird predation. The adult viceroy is also without any poison and is highly palatable. It relies for defence on its similarity to the monarch which is a much commoner species.

This is an example of 'Batesian mimicry', named after Henry Bates the English naturalist-explorer. He was the first to draw serious attention to the widespread occurrence of mimicry among animals. Bates was a remarkable field biologist. During an eleven-year expedition to the Amazon jungle he amassed a tremendous collection of animals, thousands of them new to science. In the course of his tireless work he noticed how often similar-looking butterflies, with similar habits, would prove to belong to different families when he examined them closely. Furthermore, he noted that usually only one member of such a pair belonged to a genus of which all members are brightly coloured and avoided by birds. The other had many cryptic relatives suggesting they were palatable. Bates was a believer in the new theory of evolution over which a great controversy was raging at the time. Following Darwin's reasoning, he conjectured that the mimicry had not arisen simply by chance. Instead, he believed that over a long period of time the edible species had become more and more like the distasteful one. At each of the millions of

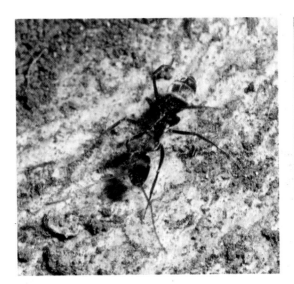

An ant-mimicking jumping spider (left) and its model (far left) from East Africa. The spider lives among the ants thereby deceiving spider-hunting birds and lizards.

tiny steps in this evolutionary process, those individuals best able to fool predators tended to live longer – and leave more offspring – than those which were less convincing. In other words, the 'fittest' survived.

Bates published his ideas in 1861, only two years after Charles Darwin's *Origin of Species*. Needless to say he came in for a share of the overall scepticism roused by evolutionary theory. Since then, however, many new facts have emerged in support of his ideas. Moreover, experiments have been performed to show exactly how, and to what extent, predators are deceived by mimics.

In a simple experiment a toad is first taught to associate the appearance of a wasp with the unpleasant experience of being stung. It is then put back onto a diet of palatable insects, but after a while hover-flies superficially similar to wasps are offered. The toad steadfastly refuses to eat them. It even shuns house-flies with painted-on yellow stripes, though it continues to eat unpainted flies. This shows that the toad associates the striped pattern itself with the unpleasant experience.

If the experiment is done differently and the naïve toad is offered hover-flies before its first encounter with a wasp, it eats them avidly. This shows two things: that hover-flies are perfectly edible and that the toad does not have an innate fear of yellow-and-black striped insects. Similar experiments have been performed with birds offered distasteful butterflies and their mimics, and with many other predator-prey situations. The results have shown conclusively that mimics do indeed fool their enemies.

Often, Batesian mimics are in the same group as their models. This is not surprising for the 'family resemblance' is a good start in the evolution of convincing deception. The viceroy butterfly had to change its coloration, and to some degree its behaviour as it evolved into a mimic of the monarch. But at least it began with the basic butterfly form.Where the model is in a different class altogether there is much more to be done. A spider with eight legs, a compact body and no antennae, has a great deal to alter before it can resemble an ant with six legs, a narrow-waisted body and prominent antennae. Yet many spiders, especially in the family Salticidae (jumping spiders), have achieved remarkable ant-mimicry. Even more distant from their models are the snake-mimicing caterpillars found in the jungles of the Old and New World tropics.

Batesian mimicry is far less common in vertebrates than in the lower orders of animals. Probably this is because there are few models available with distasteful properties, or hidden weapons. Among mammals, the best-known Batesian mimics are the forest-living squirrels in South-east Asia that resemble tree-shrews. Tree-shrews, *Tupaia*, are regarded as primitive primates by some taxonomists and as advanced insectivores by others. Like the true shrews they are inedible to most predators. Their flesh is said to be unpleasant, but more importantly they have a bitter secretion on the fur. Most tree-shrews are active during the day like the squirrels that copy them. It takes an experienced naturalist to tell some of the pairs apart. In the Malayan rainforest the greater tree-shrew, *Tupaia glis*, and the red-bellied squirrel, *Callosciurus prevostii*, constitute a pair. The lesser tree-shrew, *T. minor*, and slender little squirrel, *Sundasciurus tenuis*, make another. On the island of Borneo there are two more tree-shrew/squirrel pairs, so it seems extremely unlikely that the resemblances are coincidental.

Another case of mimicry by a mammal was suggested only recently. It concerns the resemblance between the aard-wolf, *Proteles cristatus*, and the striped hyaena, *Hyaena hyaena*. These two African savanna-dwellers share a similar colour-pattern of vertical stripes and appear to be the same size. Yet the aard-wolf is

actually a lightly built creature that hunts only small prey, and is very different from the powerfully built hyaena. The similarity in size is an illusion created by the aard-wolf's long legs and erect mane. The striped pattern on the body continues to the crest of the mane. The aard-wolf's body-weight is a mere 12–15 kg, while the hyaena weighs nearer 60 kg. Being a lightweight, it risks attack by leopards which eat small predators, including jackals and cats, as well as their more usual prey of antelopes and monkeys. But a leopard would be unlikely to attack a hyaena which is an immensely powerful animal with huge, bone-crushing jaws. Thus by resembling a hyaena, especially at night when such animals are in the open, the aard-wolf may gain a considerable measure of protection. Clearly it is not easy to test.

Left: Chrysalis of an unidentified butterfly which mimics the head of Schlegels' pit-viper, a small but deadly snake feared by birds in the Central American rain-forest.

Below: The aard-wolf, a harmless insectivore, is protected from large predators by its resemblance to the powerful striped hyena.

An even more fascinating suggestion, also made recently, concerns another pair of African mammals, the cheetah and the honey-badger. Young cheetahs are unique among cats in being completely dissimilar in pattern from their parents. Adults are spotted all over, but their kittens, till they are about ten weeks old, have a plain, silvery-white back and dark-spotted underside. At night, a young cheetah bears some resemblance to a honey-badger — the most savage animal for its size in Africa. A young cheetah away from its parents could provide an easy meal for a lion, leopard or hyaena, but with luck a predator may think it

sees the back of a honey-badger crouching in the long grass (cheetahs give birth in the wet-season) and leave it alone.

There are relatively few cases of mimicry known in birds, but one is of special interest: the mimicry of the host species' eggs by the eggs of the cuckoo. The European cuckoo, *Cuculus canorus*, like many other cuckoos, lays its eggs in the nests of small insectivorous birds. For this reason it is classified as a brood parasite. The cuckoo lays a single egg in each nest, usually before the host has completed its clutch. As she lays, the cuckoo removes one of the owner's eggs to aid the deception. On hatching, the baby cuckoo's first task is to heave the host's eggs or nestlings, one by one, out of the nest. This achieved, all the insects intended for the host's 4—6 babies go to the single youngster. The young cuckoo grows rapidly and soon outstrips its foster-parents in size.

Many small birds destroy a cuckoo's egg if they detect it in their nest. Alternatively, they may desert the nest and begin building a new one elsewhere. So cuckoos have evolved an egg-colour and pattern to match that of the host's eggs. The mimicry is defensive, though in rather a special way. The protection is needed against the host species, not a predator.

In different parts of its range the cuckoo parasitizes different hosts, and in each area it has evolved appropriately patterned eggs. The total variety is great, but each particular female lays only one colour of egg. For instance, cuckoos habitually parasitizing redstarts lay plain-blue eggs, whereas those laying in the nests of reed warblers have greenish eggs heavily blotched with brown. In both cases the

Left below: The unique pattern of cheetah kittens gives them a superficial similarity to the honey-badger (above), the most powerful creature, for its size, in Africa.

Right: A cuckoo's egg in a reed warbler's nest. The host's ability to discriminate has forced cuckoos to produce mimetic eggs.

resemblance to the hosts' eggs is striking. Cuckoos parasitizing wagtails and pipits also display good egg-mimicry, but those laying in nests of the dunnock (or hedge sparrow) produce eggs very dissimilar from those of the host. In this case the selection pressure for mimicry has not been strong. This is because dunnocks are unusually tolerant of cuckoos' eggs among their own, as experiments have shown.

The European cuckoo nestling does not need to mimic those it replaces, for the foster-parents seem willing to feed any begging youngster in their nest. But nestlings of other cuckoos, such as the Indian koel, *Eudynamis scolopacea*, look exactly like those of its foster-parents, which are crows. This mimicry is essential because the young koel does not eject its nest-mates. This may be because they are unable to, being no bigger than the young crows, or perhaps because the adult crows are two intelligent to accept one baby where there were four before. The koel is reared with the young crows and receives its share of the food.

The European cuckoo shows mimicry in another way which has been interpreted, perhaps wrongly, without reference to defence. As every experienced naturalist knows, the cuckoo can easily be mistaken for a bird of prey, such as a sparrowhawk. Apart from the similarity in pattern and form the cuckoo gives the impression of being quite a large bird, but actually it weighs no more than a starling. The size bluff depends on the cuckoo having exceptionally long feathers covering its body as well as long tail and wing feathers. The resemblance is said to trick the hosts into leaving the nest area in order to chase the 'hawk' away. It is further supposed that in the ensuing confusion the cuckoo can lay its egg in the host's nest. This hypothesis has not been substantiated by direct observations, however, and there is now evidence that many hosts pay little attention to cuckoos anyway. Also, the female cuckoo normally deposits her egg in the nest before the host's clutch is complete and before incubation has begun. So there are long

periods when the nest is being left untended by its owners and this is when the cuckoo sneaks in to lay. It is also a fact that although some other parasitic cuckoos exhibit hawk-mimicry, many do not. Indeed some of the African and Asian cuckoos are brilliantly coloured.

It seems possible that hawk-mimicry by cuckoos has been misinterpreted and is actually straightforward Batesian mimicry. By looking like a small but powerful bird-of-prey, a cuckoo probably escapes being attacked by goshawks and falcons. The deception could be especially useful when they are feeding. They tend to sit for long periods on a good vantage point waiting to pounce on any large insect seen moving below. While watching the ground in this way they make 'sitting targets' for hawks and falcons, and their hawk-like appearance may often save them. The benefits could be even greater in the cuckoo's tropical winter quarters, where birds-of-prey abound, than in their breeding areas in Europe and northern Asia.

Cold-blooded vertebrates (reptiles, amphibians and fishes) have not evolved mimicry on a large scale. They rely mainly on concealment and warning for defence, but there is one situation involving the coral snakes that is particularly interesting because it concerns a special kind of mimicry. The coral snakes are members of the Elapidae, the family of poisonous snakes which includes the cobras. Their name derives from their brilliantly coloured pattern of alternating rings of red, black and, in some species, yellow or white. Coral snakes have nothing to do with coral reefs; the majority live in the soil or leaf-litter of inland habitats. Some species are quite belligerent when they are threatened, while others are completely docile. Either way they are all potentially dangerous. Those coral snakes which feed on other snakes have a particularly potent venom to still their prey rapidly. The bold pattern of a coral snake is clearly for warning and presumably protects them from mammals that unearth them when rooting for food. Even at night, when colours cannot be discerned, the coral snake's bold pattern remains conspicuous.

Many of the coral snakes have been copied by non-poisonous snakes belonging to another family, the Colubridae. The harmless king snake, *Lampropeltis polyzona*, for example, is strikingly similar to an extremely vicious coral snake, *Erythrolamprus aesculapii*. The similarity goes beyond colour and pattern, for the king snake gives a precise imitation of the coral snake's threat-display if it is molested.

This seems like a straightforward case of Batesian mimicry and so it was considered until a herpetologist, Robert Mertens, pointed to a puzzling anomaly: in typical Batesian mimicry situations predators learn to avoid the war-

ningly-coloured model by having an unpleasant experience with it early in life. Thereafter they avoid further encounters, and this is how convincing mimics gain advantage, for they, too, are avoided. But Mertens pointed out that the coral snakes are so venomous that a would-be predator's first encounter would be its last! This, and other anomalies, induced a specialist in mimicry phenomena, Wolfgang Wickler, to suggest that another quite different form of mimicry was involved. He termed it 'Mertensian mimicry', in recognition of Merten's thought-provoking observations.

Wickler decided that the more poisonous coral snakes were not the models at all. He proposed instead that less poisonous coral snakes were the models. These are copied, on the one hand, by harmless snakes (simple Batesian mimics), and on the other by the deadly species (the 'Mertensian mimics'). According to this theory, the deadly Mertensian mimics stand to gain by not having to waste their venom on animals that disturb them but cannot serve as food.

One requirement of the theory is that forest animals disturbing one of the less venomous coral snakes (which are apparently the most numerous) are, in fact, bitten. They must then recover so that they can retain the association between the coral pattern and the bad experience. Thereafter harmless colubrid mimics and deadly species of coral snakes are both left severely alone. This is the theory, but the experiments needed to test it have not been carried out. The experiments must be performed with those forest mammal species that inhabit the same regions as coral snakes – not with the usual laboratory animals. Obviously this is no easy goal.

When the situation has been properly investigated it may well turn out that there is no need for the Mertensian mimicry concept. New World mammals that dig in the forest floor may have evolved an innate fear of the deadly coral snakes. Now they avoid them – or anything like them – from birth. If this is so, it is merely a slight variant of the usual Batesian mimicry situation. The only unusual feature is that 'knowledge' of the model's noxious qualities is instinctive rather than learned. If this is so, another oddity is easily understood: in typical Batesian mimicry the mimic is a less abundant species than the model, but, as Wickler pointed out, the mimics of coral snakes seem to be more plentiful than the models. This fact presents no problem if the

Top: The Long-nosed snake, a harmless mimic of the Western coral snake. In the figure the deadly Brazilian coral snake (family Elapidae) is compared with its mimic below, a non-poisonous member of the Colubridae.

snakes' potential enemies are born with a fear of the coral pattern already in them.

Insects provide numerous examples of mimicry, many of them baffling in their complexity. There is a good reason for this. Whereas camouflage and disguise can be effective against both poor-sighted predators and those with acute vision, mimicry can evolve only where the predators can see detail. Of all the diurnal predators birds have by far the most discerning vision. Indeed, most of them rely on sight alone for finding food since they lack the sense of smell. It is mainly birds, therefore, that have induced the evolution of so many insect mimics.

Most of them have other insects as their models. The noxious species being imitated are either palatable insects protected by a sting or poisonous bite, or highly distasteful species. The former are nearly all in one order, the Hymenoptera, which includes hornets, wasps, bees and ants. The unpalatable models, on the other hand, are a diverse assembly including grasshoppers, bugs, beetles, moths and butterflies. Some are so well protected from birds and other predators that they have been mimicked not only by palatable insects but by spiders too.

The stinging Hymenoptera are copied by a wide range of other insects — including many stingless bees and wasps. The sting is actually a modified ovipositor, part of the egg-laying apparatus. Only females have a sting, including the worker caste of social bees and wasps which are all sterile females. The males, or drones, have no sting. During the brief period they are on the wing they derive protection from being patterned like the well-armed females. Thus they are simply Batesian mimics of the opposite sex!

The hornet clearwing moth, *Sesia apiformis*, is a typical European wasp mimic. Its body is striped vividly with black and yellow like a hornet or *Vespa* wasp. The wings are clear like a wasp's instead of being covered with scales as they are in typical moths. Furthermore, the hornet clearwing is active by day, which is unusual for a moth. A similar deception is practised by the many species of bee hawk-moth, *Hemaris*, which closely resemble bumble-

Wasp mimicry has been evolved by many European insects. Above: the wasp beetle *Clytus arietis*. Right: a clearwing moth *Aegenia vespiformis*.

Left: On hatching, the nymph of the Australasian stick-insect *Extatosoma tiararum* runs about wildly like an angry tailor-ant, its front legs held aloft like antennae.

bees. They have the same fat, furry abdomen bearing a pattern of transverse stripes, short black antennae and transparent wings. The visual similarity between a bee hawk-moth and a bee is enhanced by loud buzzing created by the moth's high-frequency wing-beat. As it flies from flower to flower in search of nectar it looks and sounds like one of the many bumble-bees doing the same.

Many of the two-winged flies (Diptera) are also wasp- or bee-mimics. They include members of many different families, among them horse-flies, bee-flies, hover-flies and robber-flies. Some are smooth, slender and striped with vivid yellow and black, like wasps. Others have a fat, hairy body and less intense stripes, like bees. The hover-fly, *Volucella bombylans*, is especially interesting for it exists in several forms, each mimicking a different species of bumble-bee. This example is complicated by the fact that the fly in question lays its eggs in bumble-bee nests. Consequently its bumble-bee mimicry has often been ascribed to deceiving the bees into letting it into their nests. This is a somewhat facile explanation, however, for insects recognize each other more by smell and touch than by close visual scrutiny. It is more probable that *Volucella bombylans* resembles the bees so that ever-watchful birds do not recognize it as prey — especially when it is hanging around the bees' nests waiting for an opportunity to enter.

When looking at the great variety of flies on summer flowers, one sees many convincing mimics of the familiar social wasps, honey-bees and bumble-bees. Others look far less convincing

mimics, but often this is because their models are not familiar to us. They are actually good mimics of potter wasps, ichneumon wasps and solitary bees. We do not think of these as unpleasant insects because they do not willingly sting humans. Nonetheless they may be well known to birds that have tried to eat them and have been stung on the tongue.

There are beetles also, in Europe, that resemble wasps and bees. The wasp beetle, *Clytus arietus*, has a very bold pattern of black and yellow stripes and appropriately wasp-like behaviour. It runs about actively in the sunshine, constantly moving its antennae and buzzing when disturbed. The wasp beetle makes it buzz by rubbing hard parts of the body together so that they vibrate.

Black and yellow transverse stripes are the hallmark of wasps, hornets and bees in the temperate zone and their mimics living in the same regions have this same pattern. But in many tropical habitats the most feared stinging Hymenoptera are completely black, or black with an orange tail. Their mimics are coloured accordingly.

Ant-mimicry differs somewhat from wasp- or bee-mimicry in that the ants serving as models are the wingless worker caste, not the winged sexuals which are rarely seen. This requires that any mimics are also wingless, or if they do have wings they must keep them well concealed. Another difference is that ants are not brightly coloured nor strongly patterned like wasps and bees. They are recognizable for what they are because of their characteristic shape and intense activity. A good ant-mimic must have the same narrowly-waisted abdomen, long legs and elbowed antennae. It must also move about constantly and show aggression when confronted. An 'ant' that moves sluggishly and retreats when approached would hardly fool an experienced bird or lizard.

An exemplary ant-mimic is the first instar nymph of a stick-insect from New Guinea, *Extatosoma tiararum*. The typically long legs, uniform coloration and small size of a young stick insect are well suited to the task, but the extremely short antennae and long, straight abdomen are quite unlike those of an ant. It surmounts the second of these shortcomings by tightly curling its long abdomen. Seen from above, the abdomen appears to be short and rounded, and connected to the thorax by a narrow waist.

Unusually for a stick-insect this creature races around with great agility and if threatened moves quickly towards the source of danger, raises its front pair of legs and moves them about wildly. The overall impression is of a belligerent tailor-ant racing around with antennae waving to defend not only itself but all the other members of the colony. Tailor ants, *Oecophylla*, are extremely common in many habitats in the Old World tropics. They are shunned by man and beast alike for they defend the nests built among the foliage with great zeal, and their bite is painful.

Other insects have evolved mimicry of ants. The bugs and grasshoppers which do so have no waist; instead the illusion of a constriction is given by an appropriate pattern. Quite simply, the waisted abdomen is 'painted on' in black or some other dark colour, while the remainder of the mimic's body is similar in colour to the background. A most bizarre ant-mimic is found in South America where its model, the leaf-cutting ant, *Atta*, is plentiful. Leaf-cutters can be seen almost everywhere as they file back to the nest, each carrying a large piece of freshly-cut leaf over its back. The entire combination – ant plus leaf – is simulated by a tree-hopper (Membracid). It has a brown head, legs and abdomen, and a huge flat extension to its thorax projecting vertically over the body. The extension is coloured bright-green, just like a freshly-cut piece of leaf. Only a leaf-cutting ant would know the difference!

Below: The older *Extatosoma* nymph, too large to resemble an ant, becomes scorpion-like and moves deliberately like its model.

Many of the spiders resembling ants belong to the family Salticidae. One large genus is *Myrmarachne*, which literally means 'ant-spider'. Salticids are jumping-spiders, so called because they leap on to small flies and other insects which they stalk with the aid of their huge eyes. Constant diurnal activity and small size befits them for ant-mimicry, while the required waisted outline is easily simulated by a small modification of the spider's abdomen. But since arachnids are devoid of antennae these have to be simulated. The spider lifts its first pair of legs whenever it stops, and vibrates them like the antennae of a questing ant. In some species the front legs are held in a crooked position so that they look even more like an ant's elbowed antennae. Ant-mimicry has evolved separately in at least nine families of spiders, providing yet another example of convergent evolution. Further testimony to its efficiency in protecting spiders is given by the fact that in some regions one in every hundred species is an ant-mimic.

One ant-mimic which merits fuller description is *Bucranium*, a crab spider living in the Amazonian forests. After feeding on a large black ant is raises the corpse and carries it over itself as it walks about in search of another victim. It is not clear whether the subterfuge assists it in catching the ants, but it certainly confers protection from birds. The combination of the ant's head and body with the spider's legs, seen from above, gives the appearance of a live ant hurrying along the trail.

Among the insects which make good models are many beetles shunned by birds because of their acrid secretions, or evil flavour. They include literally thousands of species in many different families. Best known are the ladybirds (Coccinellidae), leaf beetles (Chrysomelidae), tiger beetles (Cicindellidae) and blister beetles.

Their brilliant warning patterns of spots, stripes or iridescent sheen have been closely simulated by palatable moths, cockroaches, grasshoppers and plant-bugs, as well as by edible beetles in other families. Sometimes the copy is so accurate that even the specialists are deceived. A grasshopper, *Condylodera*, which closely mimics a tiger-beetle, *Tricondyla*, was first discovered in a museum collection in Borneo, where for years it had gone undetected, pinned among the beetles.

It is the butterflies, however, that display the most complicated mimetic associations. This is presumably because they are completely diurnal and although relatively difficult to catch on the wing they are vulnerable to bird attack when at rest or feeding. By contrast, there are only a few mimetic moths and these fly by day, unlike most of their kind.

African monarch butterflies, like their American relatives, are distasteful to birds. Consequently they have been imitated by numerous palatable species. The 'beautiful monarch', *Danaus formosa*, a high-flying species of forested areas, is the model for one of the swallowtails, *Papilio rex*, which also lives in the tree canopy. Similarly the 'blue monarch', *Danaus limniace*, is mimicked by *Graphium leonidas*, another swallowtail. In these examples the model is copied by both sexes of the swallowtail, but this is not always the case with butterfly mimics.

The well-named 'mocker swallowtail', *Papilio dardanus*, has a wide distribution in Africa. Everywhere the male is a typical swallowtail: a lovely cream-and-black insect with long 'tails' extending from its hind wings. In one geographical race *P.d. meriones* (taxonomy shorthand for *Papilio dardanus meriones*), confined to the island of Madagascar, the female is a typical swallowtail like the male. But on the mainland the females have lost the swallowtail look and become mimics of distasteful monarch butterflies. Males and females are therefore completely dissimilar to the extent that a mating pair looks very odd indeed. Why, in this swallowtail, the males have not also become mimics is difficult to understand. But this is by no means the only complication, for the females living in different regions have copied different models. For example, typical females of *P.d. dardanus* in West Africa resemble the boldly patterned black-and-white *Amauris niavius* while females of the East African *P.d. tibullus* mimic *Amauris albimaculata*, a more finely marked monarch with yellow patches on the hind-wings.

This is in no measure the full story and in some ways the above is over-simplified. Those who would like to delve more into the situation, involving as it does seven geographical races of *Papilio dardanus* (many with different 'morphs') as well as other species of butterflies, are referred to Wolfgang Wickler's excellent *Mimicry in Plants and Animals*. Vastly more complex mimetic associations, involving hundreds of butterfly species belonging to many different families, have evolved in the jungles of tropical America. It will take several centuries for entomologists to unravel the whole of this baffling tangle of species, races and morphs.

As with any other defence strategy, Batesian mimicry is not without its failings. To be really effective the mimic species must be less common than the model. If it is more plentiful, predators will encounter and consume many mimics before they first come across one of the unpalatable models. The possible consequences of this would vary from case to case. If the eventual bad experience with the model is extremely unpleasant, the predator would subsequently avoid the mimics, too, no matter how abundant they are. But if the model is merely distasteful and can be spat out, it might

Similarity between members of the families Danaidae and Acraeidae exemplifies 'Müllerian' mimicry, since all are toxic. The representatives of the two families (right) are all palatable 'Batesian' mimics of the toxic species.
In the case of *Papilio dardanus* the male is a typical, non-mimetic, swallowtail. The females are mimics and occur in many forms including the two shown, one a mimic of *Danaus chrysippus*, the other of *Amauris niavius*.

*Danaus chrysippus*

*Amauris niavius*

DANAIDAE

*Acraea encedon*

*Bemanistes epaea*

ACRAEIDAE

*Hypolimnas misippus* ♀

*Hypolimnas dubius* ♀

NYMPHALIDAE

PAPILIONIDAE *Papilio dardanus*

♀ ♂ ♀

then pay a predator to carry on regardless. The risk of biting an occasional bitter-tasting item would be worth taking as long as the majority prove edible. In this hypothetical situation, where the palatable mimic greatly outnumbers the distasteful model, it would not take a bird long to distinguish between model and mimic. Experiments with pigeons and other birds have shown what subtle differences between items they can detect in choice tests where only correct decisions are rewarded.

As long as the model species predominates, the problem of discrimination by predators is unlikely to arise, but animal populations are not static. The numbers of the mimic and model species are regulated by different sets of factors. If the model suddenly increases greatly in abundance this can only benefit the mimic. But if the mimic does extremely well one year while its model fares badly the mimic's defence strategy might begin to fail. Some mimics mitigate this hazard by coming on the scene, as it were, later than the model. An instance of this, suggested recently by entomologist Miriam Rothschild, concerns two British butterflies, the brimstone, *Gonepteryx rhamni*, and cabbage white, *Pieris brassicae*. The cabbage white is known to be unpalatable to birds; it contains sinigrin and mustard-oils derived from cabbage and other larval food-plants. The brimstone, on the other hand, is palatable. At rest it is protected by its disguise as a leaf; the closed wings are pale green and marked with a leaf-vein pattern and mould spots. When it flies the brimstone becomes vulnerable and for this reason the female, which appears white on the wing, mimics the cabbage white. Now the numbers of these two butterflies vary greatly from year to year, but always brimstones appear later in the summer than do their models. So no matter how the relative numbers turn out, naïve young birds will have had memorable experiences with cabbage whites, causing them to avoid large white butterflies, before they see their first brimstone. One problem with this interesting theory is that some of the brimstones hibernate and appear again the following spring – before there are cabbage whites about. Miriam Rothschild has anticipated this by explaining that when the brimstones are flying in early spring there are, as yet, no newly-independent young birds to try them. And older birds retain memory of the cabbage white from previous years.

A more intractable constraint on a mimic than its abundance is its geographical distribution. For it can never extend its range beyond that of its model. Also if the range of the model is decreased for any reason those parts of the mimic population left behind are doomed.

When Bates formulated his ideas on mimicry he was at a loss to explain why although most

Above: Hibernating queen wasp. Because of their stings social wasps are avoided by most birds and their characteristic pattern is imitated by many other insects.

mimics were palatable there were some that were obviously distasteful like their models. Several South American butterflies belonging to different families, which he knew to be unpalatable to birds, shared a common distinctive pattern. An explanation was offered by the German-born Brazilian naturalist Fritz Müller who, like Bates, had extensive knowledge of Amazonian butterflies. Müller suggested that by sharing the same pattern each species profits from predators' early experiences with one of the others. They all have unpleasant properties that would make a bird drop them quickly, but since even a foiled attack might lead to damage, it is best avoided. One cannot talk about 'models' and 'mimics' in this situation, for Müllerian mimics are mimicking one another.

In Europe and North America there are mimetic associations of this kind involving many species of hornets, wasps and bees. All have transverse black and yellow stripes, and a young bird experiencing the sting of any one species is likely to leave the others alone too.

In any particular area the number of similar species may add up to hundreds and unravelling the situation can be difficult. New mimetic associations, some of extraordinary complexity, are being discovered all the time, especially in the tropics. It is not always possible to distinguish the two kinds of mimics. A pair of warningly-coloured species may both be repulsive to a number of predators, in which case they should be regarded as Müllerian mimics. But to some other predator one of them may be palatable, in which case it benefits as a Batesian mimic of the unpalatable one. In other words an individual can be a Batesian or a Müllerian mimic, depending on the particular predator by which it is confronted.

This text is concerned essentially with animals and their defensive adaptations. Nevertheless it is useful to look briefly at some plants in the context of mimicry, if only to emphasize that some principles cut across the boundary between the two kingdoms. It would be out of place to dwell on plants that mimic other plants — though there are numerous fascinating examples of this — but a situation where a plant mimics an animal certainly merits description.

The huge order of orchids includes a large number that offer a reward of nectar to insects and other animals relied upon for cross-pollination.

In this respect they are like many other kinds of flowering plants. But there are other orchids that use insects to transfer their pollen without wasting any resources on the production of a nectar reward. Instead, they trick insects into making contact with the flower by mimicking the insect's own kind. The bee orchid, *Ophrys apifera*, has a flower which looks, even to our eyes, just like a hovering bumble-bee. The lower part of the flower simulates the bee's fat, furry abdomen and bears a pattern of transverse stripes like the insect's. Two other floral parts project sideways in imitation of the insect's wings; long hairs on these projections give the impression of the blur created by a rapid wing-beat.

Evidently the ingenious mimicry can fool a male bumble-bee to the extent that it attempts to copulate with the flower! In the process it gets pollen stuck to itself, and transfers to the flower any pollen picked up from flowers 'visited' elsewhere. Other orchids are known to mimic solitary bees, wasps and flies in order to induce pseudocopulation, as it is called.

Below: Bumble-bees also have their mimics, like the bee hawk-moth from southern Europe.

# Patterns that Startle and Deflect

Animals relying on camouflage or disguise as their main line of defence often have a second line to fall back on if a predator comes too close. This can happen, for instance, when a bird searching among foliage accidentally disturbs a creature it had not previously noticed, or when a cryptic individual chooses a background against which it is relatively conspicuous. In a later section defensive routines will be described in some detail. Here it is enough to note that the secondary defensive ploy is often some kind of startling display. It may be a mock threat made by a harmless species, or a very real threat by a cryptic animal with a potent weapon. Whether real or false, a threat-display is intended to frighten the enemy away altogether. Other startling displays are effective only momentarily; in other words they have shock-value, but no real meaning. If the predator has already seized its victim the display hopefully causes it to let go again, providing an opportunity for speedy escape. As a peculiar variant on this theme, some animals feign death as soon as they are molested. In this event the adversary is not exactly startled, but if it can react only to moving prey — as with many predatory insects, frogs and lizards — its feeding response is stilled. After some minutes it loses interest, and goes away. A few death-feigners perform so convincingly that they look *and* *smell* like a long-dead corpse. They aim to dissuade bird or mammal predators, most of which eat only freshly-killed prey.

Sudden displays of this kind are not the prerogative of cryptic animals alone. An aposematic species may also employ a warning display to emphasize the point that it should not be trifled with. If its weapon is visible, the display may be a flourishing of the weapon itself. Less obvious noxious attributes such as poisons in the body can also be underlined by displays accompanied by production of an acrid or foul-smelling odour. This can be a good deterrent against mammalian predators. Against birds, the warning display must be visually spectacular.

Deflection, the other main topic in this section, refers to patterns or activities that cause a predator to aim wrongly. It either misses its target altogether or is persuaded to aim at the least vulnerable part of its intended victim's body. In the latter event it ends up with a mouthful of fur, a piece of wing or a detached tail — none of them items with a high nutritive value.

Beginning with displays that aim to frighten, the first generalization to make is that they must be dramatic and quick. Any colours or patterns suddenly revealed must contrast strongly with those visible beforehand if they are to have maximum impact.

Numerous morphological adaptations are known that permit a sudden show of colour. The crested rat, *Lophiomys ibeanus*, simply parts the hair on its back to reveal black and white stripes. The Australian frilled lizard, *Chlamydosaurus kingi*, spreads a brightly coloured ruff as it stands open-mouthed to face an attacker. In a similar fashion many fishes spread their fins to reveal previously hidden colours. The European fire-bellied toad,

The *Thecla* butterfly of tropical America is a master of bluff. Bold lines converge on its false eyes and antennae-like filaments on the hind-wings, giving the impression of a watchful, raised head. But if disturbed, the butterfly takes off in the opposite direction.

Left: A tokay gecko from S.E. Asia gapes widely in mock threat. The jaws are ineffectual for defence against large animals.

*Bombina bombina*, turns on its back when alarmed and lies scarlet belly uppermost, a technique used also by Australian toadlets.

Insects often have large areas of colour on their hind-wings. At rest they are covered by the fore-wings which may bear a cryptic pattern, but by raising the first pair of wings the bright colour patches can be prominently displayed. Alternatively the colour may show only during flight, as with the many underwing moths in the genus *Catocala* which have dull-brown fore-wings and vivid red, yellow or blue hind-wings.

Looking at all these colourful displays one is bound to wonder how effective they really are at saving the animals' lives. Experiments have been carried out with underwing moths in the U.S.A. to determine how the brightly coloured hind-wings give protection. Often a bird disturbing a moth resting on tree bark was so frightened by the sudden appearance of the strong pattern when the insect took to flight that it failed even to follow. Those that did set off in pursuit often failed to find the moth again. This is because in the bird's mind it was following a yellow, red or blue insect and this mysteriously disappeared. When the moth landed its hind-wings were quickly folded and covered by the pair in front which are cryptically coloured. To aid this confusion underwing moths typically fly fast and far through the trees before they land.

A third attribute of the colourful hind-wings is to deflect the beak snap of the bird – if it comes – away from the vulnerable head and body. Large numbers of *Catocala* moths were collected with light-traps and examined for the distinctive triangular marks left by birds' beaks. Such marks indicate that the insects had escaped sometime during the preceding days. The percentage of marks across the hind-wings was greater than would be expected by chance alone. This suggests that birds often aim at what they see most clearly when chasing an underwing – the colourful hind-wings. How often the moths escape after being grabbed this way we cannot know. Obviously some do, or they wouldn't turn up subsequently in light-traps.

Pigments used in startling displays can sometimes be in liquid form. The sepia ink squirted into the water by a frightened squid probably has shock value as well as providing a smoke-screen behind which the squid can escape. Deep-sea species eject a luminous cloud and this must surely startle fish predators. Another fluid-show of colour is exhibited by insects that regurgitate their green stomach contents. Others exude a black liquid from the mouth and some produce red blood by voluntary internal bleeding. One large African mantis drips the blood from its mouth as part of an impressive display which also involves an illusory size-increase obtained by spreading the wings.

Patterns that resemble eyes can be much more effective than simple patches of colour — especially if they depict huge staring eyes of a vertebrate. Eye simulation is achieved by concentric rings of pale colour round a dark centre which represents the pupil. An illusion of roundness may be given by a highlight spot to one side of the 'pupil'. This simulates the shine on a convex surface. The European eyed hawk-moth, *Smerinthus ocellata*, relies on disguise as its first line of defence as it rests in dense herbage. But if the disguise is detected the large fore-wings are snapped forward to reveal two large 'eyes', surrounded by a shocking-pink colour, on the hind-wings which startle the predator. False eyes of this kind, hidden away on the hind-wings, have been evolved by other cryptic insects, including grasshoppers, bugs and mantises. Caterpillars like that of the elephant hawk-moth, *Deilephila elpenor*, have two sets of false eyes on the thorax. When alarmed, the true head with its tiny ocelli (simple eyes) is withdrawn and the false eyes suddenly enlarge greatly in size as the thorax expands, simulating a large-headed monster. The adaptation has been taken further by certain tropical hawk-moth caterpillars that have an entire snake's head design on the thorax. The caterpillar of *Leucorhampha* is

normally cryptic, but if disturbed rears up from the branch on which it rests, waving its expanded thorax from side to side and striking viciously. The snake head is depicted in fine detail with all the scales and two shining eyes. The related *Pholus labruscae* larva has eyes that appear to wink as the snake display is performed.

One of the oddest of all insects is the South American lantern bug, *Fulgora lanternaria*. A grotesque, hollow false head, complete with tooth-filled jaws and protruding eyes, in front

Above: The threat of this African mantis is made more menacing by 'blood' dripping from the mouth.

Left: Increase in apparent body size is a common bluff tactic. The huge neck frill and open jaws of the Australian frilled lizard give it a frightful appearance.

of the real head, produces a constantly frightful appearance. And if the 'alligator look' fails to deter an inquisitive predator a pair of large eye spots on the hind-wings are suddenly brought into prominence.

Huge intimidating false eyes have been evolved also by some of the higher animals. In Brazil there is a toad, *Physalaemus nattereri*, with enormous false eyes on the lower part of its back. When disturbed it turns away from the source of danger, puts its head down and confronts the intruder with the huge staring eyes on its rump. An unidentified frog found in the lowland forest of Costa Rica has a similar display, but the 'eyes' are on the hind legs.

There is even a bird that could truly be called 'four-eyed'. The pearl-spotted owlet, *Glaucidium perlatum*, a diminutive insectivore common in the African savanna, has a pair of false eyes made of small feathers on the back of its head. They are a little larger than the true eyes and can be revealed voluntarily by subtle adjustments of the plumage. My own first encounter with one of these little owls, in Nigeria, may point to the true value of the extra

pair of eyes. I was first guided to the owlet by the clamour of sunbirds and warblers mobbing it as it rested in an Acacia bush. It was unafraid of the small birds darting around its head and to my surprise allowed me to approach closer and closer. All the time it was apparently looking me straight in the face with its staring black eyes. Then suddenly, when I was only a few feet away, the owlet must have heard the movement and turned its head round. The sight of another face looking at me, this time with orange eyes, certainly took me aback. Meanwhile the owlet flew off rapidly to some distant trees. In this episode, and perhaps always, the value of the false eyes was not their startling effect but the shock-value they allowed the real eyes to have when the head was turned to face the 'attacker' coming from behind.

Experimental proof of the value of false eyes for frightening predators is difficult to establish, but there is sufficient anecdotal evidence that birds in particular are discouraged by displays incorporating the sudden show of large false eyes. Dr. H. B. D. Kettlewell, the entomologist

Above: Normally the European eyed hawk-moth can rely on disguise as a crumpled dead leaf to escape detection during the day

... but if disturbed it flicks its fore-wings forward to reveal huge staring eyes on a blushing-pink 'face'.

who elucidated the industrial melanism phenomenon in the peppered moth, carried out some illuminating tests during a visit to Brazil. From the clouds of moths assembling at night round a light he collected a species of *Antomeris*, which has prominent false eyes on its hind-wings. Next morning the moths were placed in the open for birds to take. Surprisingly they were all seized so rapidly they had no time to perform their defensive eye-flashing display, but when a bird put one of them on the ground ready to tear it apart the false eyes were displayed and as the bird paused in its attack the moth made its escape.

An eye-flashing display can be effective only if the predator has not recently experienced the same display performed by other individuals. Caged birds offered a succession of palatable insects with the eye pattern soon cease to be intimidated — they become 'habituated' — and eats them with alacrity.

Given this possibility that predators may habituate, one might expect that animals relying on this stratagem would all live at low density. In the main this is true, but there are

a few anomalies: the European peacock butter-fly, *Vanessa io*, for instance, has two pairs of false eyes on the upper surfaces of the fore- and hind-wings, while underneath it is uniformly dark. When resting the peacock looks like a dead leaf and it is known to flick its wings open to reveal the eyes when molested by a bird. Peacock butterflies are often too numerous for the performance to retain its shock value, but the eyes are still of use for they have a second function — as 'deflection-marks'. That is, they direct attack towards the tips of the wings which can stand a certain amount of damage without the insect suffering real harm. This other role of eye-spots will be discussed in detail later.

The displays described so far may well intimidate predators, but they are somewhat ambiguous since they give no indication that the animal can actively defend itself. Indeed many of them cannot, their terrifying appearance being pure bluff. The threat-displays of well protected animals are in a different category. Their weapons are prominently exhibited so as to leave their enemies with no doubt

Right: The false eyes on the thighs of this Costa Rican frog may startle pursuing predators and allow time for escape.

concerning their ability to defend themselves. The ultimate aim is to avoid conflict, for fighting might be harmful – even to the winner. Often the weapon is made to stand out by the association with bright colours. The caterpillars of one Lymantriid moth from Africa have bunches of irritant hairs which can be deployed if their camouflage fails. The tufts of hairs are coloured deep-red and are normally kept hidden in pockets of skin.

To humans the most terrifying threat displays are those of venomous snakes. The ritualized threatening behaviour of the cobra is passed off as a hypnotic dance by 'snake-charmers' who pretend that the deaf snakes are responding to their music. Cobras react to the presence of any large animal by rearing up from the ground with the neck expanded into a 'hood', while making stabbing movements towards the intruder. To an animal the hood must look like a huge head and at least one cobra has false eyes. This is the Indian cobra, *Naja naja*, with the familiar 'spectacles' pattern on its stretched hood. The African black-necked cobra, *N. nigricollis*, displays bands of black or crimson on its hood. It is also known as the spitting cobra, for if the threat display is not heeded it projects venom through hollow fangs into the eyes of the intruder. The venom can be spat accurately a distance of two metres or more. The effect on a large mammal is temporary blindness and intense pain. When the spitting

cobra is hunting it does not spit but injects the venom into its prey like any other poisonous snake.

The most impressive display of all is that of the hamadryad, or king cobra, *Naja hannah*, of tropical Asia. This is the largest of all poisonous snakes, growing to four metres in length. The bite is lethal, yet it is quite inoffensive to humans and feeds mostly on other snakes. The hamadryad seeks only to be left in peace and if approached indulges in a terrifying display rather than strike. When fully reared up its head is over a metre above the ground and the spread hood displays a brilliant orange colour. This is normally sufficient to scare off all but the most foolhardy of creatures. The highly cryptic rattlesnakes of western North America threaten mainly by sound. The many species of 'rattlers', *Crotalus*, and 'pygmy rattlers', *Sistrurus*, have evolved a unique technique for making a loud buzzing noise: at each shedding of the skin, three or four times a year, the old skin is retained as a hollow shell by a 'button' at the tip of the tail. Gradually a series of these shells accumulates to form the rattle. They resonate loudly when the alarmed reptile vibrates its tail. A rattlesnake's buzz is known and feared by large mammals such as bison, especially if they have previously experienced a painful bite. Though lethal to man the dose of venom would not normally kill such large creatures. Rattlesnakes have evolved this audible threat to avoid being crushed by hefty animals with poor eyesight. The mojave rattlesnake, *Crotalus scutulatus*, has a conspicuous black-and-white banded tail in addition to the rattle.

Monkeys, dogs, cats and many other mammals bare their teeth, which stand out prominently against the pink or black gums. The effect is enhanced by accompanying growls or hissing noises and menacing movements suggestive of attack. Donkeys, zebras and horses also show their teeth, but cows and other antelopes do not. Instead they show off their sharp horns by appropriate head movements. The horns frequently contrast in colour with the surrounding skin and their tips are often darker or lighter than the rest to emphasize their sharpness.

When the weapons are not on the head, threat postures can be odd indeed. An African porcupine will face away from an aggressor, rattling the vicious black-and-white spines in its face. The spotted skunk, *Spilogale putorius*, whose weapon is a well-directed shot from its anal glands, does a handstand when confronted with an enemy. In this position the skunk's back, patterned with black and white stripes and spots, is prominently displayed, and the fluffy white tail is shown off to full advan-

tage. The vertical stance, even though inverted, also makes it look considerably bigger. A spotted skunk may even walk on its 'hands' towards an intruder, finally squirting the stinking defensive liquid into its face.

Another animal with a rear-borne weapon is the scorpion. When threatening, a scorpion faces the aggressor with the pincer-like 'chelae' raised and the tail arched over ready to strike forward. Some species also make a characteristic rasping noise. No bright colours are involved in the display. Scorpions spend the day in crevices and are therefore at risk mainly at night when they are in the open. Their unmistakable body-form and movements, reinforced often by sound, is sufficient advertisement, since most nocturnal predators cannot see colours anyway.

Just as sound may combine with colour in frightening and threat-displays, sudden odours may also serve to accentuate a visual display. The European puss moth caterpillar normally relies on disruptive patterning for concealment, but it performs an elaborate threat-display when molested. Red patches appear on the head, and long red filaments are extruded from the forks of its tail. At the same time a drop of pungent-smelling formic acid is secreted as a warning that it can produce a lot more acid if required.

Below: Like most snakes the Mojave rattlesnake does not use its venom willingly for defence. Its banded tail and audible rattle give conspicuous warning.

Overleaf: Burchell's zebras feeding. In thick vegetation the stripes have a disruptive effect, aiding concealment. But in the open, the dazzling pattern may confuse predators chasing the tightly-packed herd.

Swallow-tail caterpillars in the genus *Papilio* evert a similar red fleshy filament when alarmed, but it appears from behind the head. This strange organ, called an 'osmaterium', carries volatile chemicals with a disagreeable and often irritant odour. After flying about for a few seconds to disseminate the smell it is withdrawn again into its pocket.

Deflection marks have been evolved by many animals to direct attacks from the head and other vital parts and towards the tail or other extremity. In their simplest form they are merely prominent marks, but when highly evolved they often take the form of eye-spots. However, these are not the large intimidating eye-spots associated with threat-displays which are normally kept hidden till required; they are often small and may be permanently on show. Many butterflies have several spots close to the margins of the wings. Often they are on the undersides so that they show when the insect is feeding or at rest. It might seem fanciful to believe that birds would be deceived by such a simple trick, yet there is no doubt that they can be. It is commonplace to see butterflies with prominent triangular marks on their wings or a piece cut out of the wing where a bird has grabbed the insect and subsequently lost it. Such marks are more often concentrated near the eye-spots than would be the case if chance alone dictated where the butterfly was seized.

Apart from deflecting an attack towards less vulnerable organs, an eye-like pattern towards the rear of the body may prevent damage altogether. Predators specializing in fast-reacting prey do not indulge in long, energy-consuming chases. If they did the smaller, nimbler prey would often escape. Instead the victim is approached stealthily, or the predator waits motionless till it comes close enough for a strike. But no matter how fast the final dash or lunge, the reaction of the prey may be even faster, so that it is already on its escape flight before the predator reaches it. The experienced predator anticipates this and directs its attack in front of the sitting target. But to do this it must know which is the head-end and this is where deception can be made. By having a conspicuous false eye — or better still a complete false head — at the posterior end of the body, the predator is tricked into aiming just *behind* its target. The intended victim then escapes in the opposite direction. The deception works even better if the real head is hidden, and better still if it is disguised as a tail!

It is possible to find species that show all stages in this deflection strategy, from a crude eye-pattern on the tail to a complete false head so convincing that even the human observer is deceived. Many Lycaenid butterflies (blues and hairstreaks) have evolved false head patterns and appropriate behaviour to further

Right: Fat-tailed scorpion *Androctonus*, a deadly species from the Sahara Desert. Scorpions are active only at night when bright colours would serve no purpose. Their distinctive shape is easily recognized, however.

Right: Clicker butterflies of the American rain-forests are well camouflaged when at rest on tree bark. If threatened they escape with a downward swoop, clicking their wings audibly.

the deceit. When seen in flight, African *Iolaus* butterflies have bright-blue wings. The only unusual feature is a pair of long filaments projecting from the hind-wings, but when the butterfly lands it looks very different. Only the undersides of the closed wings can be seen and the general tone is very pale. Close to the base of each tail filament is a conspicuous eye-spot, usually orange encircled with black. Some species have two or three such eye-spots close together. One has to watch one of these butterflies feeding to appreciate the perfection of its adaptive coloration. It moves slowly on the flower head with the head lowered and the antennae still. The hind-wings are continuously moved up and down so that the filaments move like inquisitive antennae. A shorter pair of filaments below the false antennae simulate

the palps (sensory organs near the mouth). Presumably a bird sees a butterfly with its head raised, eyes staring and antennae quivering, ready to fly off at the slightest sign of danger. If the bird makes a dive, aiming just in front of the 'head', it is likely to miss completely as the insect darts off in the other direction.

Similar false-head patterns can be seen on Lycaenid butterflies the world over. Some members of the genus *Thecla*, native to tropical America, have a bold, striped pattern on the underwings. The stripes radiate from the false head, furthering the illusion in an uncanny way. For convergent lines guide the eye towards the point of convergence; a principle every artist learns as a student. Moreover, the veins on a butterfly's wings radiate from the bases of the wings, close to the head, and markings usually follow the lines of these veins. The stripes on the wings of *Thecla* run at right angles to the main veins, making the pattern even more confusing. There are also plant bugs with a false head, complete with eyes and 'antennae', at the rear of the folded wings.

Exactly the same defensive adaptation has been evolved by certain fishes. The forceps fish, *Forcipiger longirostris*, is a particularly fine example. It has a broad body and a long fine snout with which it extracts food from crevices among the corals. The general body colour is bright yellow, but the head is white below and black above. The eye is optically obliterated at the junction of the black-and-white halves. The caudal (tail) fin is rendered invisible by transparency but the anal fin just below is yellow like the body. It is also shaped like a typical fish's head, and bears a prominent black eye-spot. When the forceps fish is gently probing among the corals, always ready to dart into a hole if threatened, it appears to be facing outwards. The deception may not look as convincing to us as that of the butterflies described above, but it will confuse the relatively simple-minded predators of the sea. Butterflies have to foil attacks by birds, which are much more perceptive than fishes.

Interestingly, although fishes normally have their deflective eye-spots on the fins, they are almost invariably on the dorsal, anal or caudal fins. These are unpaired fins in the median line of the body. The other fins, pectoral and pelvic, are in pairs. Though often small they are essential for proper balance and manoeuvrability. A piece taken out of one of the paired fins could be a serious disability; hence the lack of eye-spots on them.

Good examples of eye-spots on fishes may be seen in aquarium collections. A close relative of the forceps fish is the long-nosed butterflyfish, *Chelmon rostratus*. Basically it is white with vertical stripes of orange and black, a pattern that offers concealment among long stems. One stripe runs through the eye concealing it most effectively, while on the dorsal fin is a huge eye-spot. The scissors-tail rasbora, *Rasbora trilineata*, of Malaysia has two eye-spots, one on each lobe of the deeply forked tail. As fish fanciers may have noticed, the rasbora's spots are extremely prominent when it is actively swimming, but they disappear almost completely when it rests on the bottom and can rely on camouflage. The red-nosed tetra, *Hemigrammus rhodostomus*, from the Amazon goes one better by having three eye-spots on

Left: The copperband butterfly-fish *Chelmon rostratus* has its eye obliterated by a vertical eye-stripe and a head-like pattern on its tail. Predatory fishes see it facing out from the coral when it is actually facing inwards — ready to dart into a crevice. The cleaner shrimp is a beneficial cohabitant of the reef.

the tail, one at the base of the caudal fin and one on each lobe.

Reptiles and mammals often use the tail to deflect attack away from the head. Many lizards have brilliantly coloured blue or red tails, contrasting strongly with the dull-brown body. Some also have the habit of moving the tail when standing still. Presumably this increases the possibility that an attack, if it comes, will be directed to the tail. The zebra-tailed lizard, *Callisaurus draconoides*, of North America, has the conspicuous striped pattern from which it gets its name confined to the under-surface. When molested it runs off with its waving tail raised to show the stripes. In association with such behaviour many lizards have evolved 'autotomy', which allows them to break off the tail voluntarily. The isolated tail wriggles vigorously for many minutes, captivating the attacker and giving time for the lizard's escape.

Snakes, too, may have deflective tail markings, and quite a few have a false-head pattern. The so-called two-headed snake, *Cylindrophis rufus*, of Malaya, hides its real head when molested and raises its tail, which is shaped like the head end (hence the name). The underside of the tail is vivid red, adding a menacing touch to the display. The African sand boa, *Eryx muelleri*, also hides its head under coils of its body when threatened and lifts the head-like tail. Indeed, instances of such deceptive behaviour by snakes are known world-wide. Since snakes do not practise autotomy the value of directing attack to the tail must presumably lie in the opportunity it allows for defensive use of the fangs. Snake-predators aim to prevent the fangs being used by seizing the head. If the tail is mistakenly seized instead, the snake has an opportunity to strike its attacker. Like other constrictors, the sand boa cited above is not a venomous snake; it can, however, inflict terrible wounds by sideways slashing movements of the teeth.

For mammals, deflection is not a simple matter, for they have no appendages with which they can dispense, or regrow. What they can afford to lose is hair. The Saharan jerboa and some other cryptically coloured desert rodents have a conspicuous black-and-white tuft at the tip of their otherwise bare tails. Though it may serve also for social signalling it seems likely that the jerboa's 'flag' will be grabbed at by pursuing jackals and foxes. In such an event their only reward would be a mouthful of the loose hair.

Below: The colourful tail of the *Ameiva* lizard directs attack to the least vulnerable part of its body. Many lizards can detach the tail which continues to wriggle for several minutes — holding the attacker's attention while the lizard escapes.

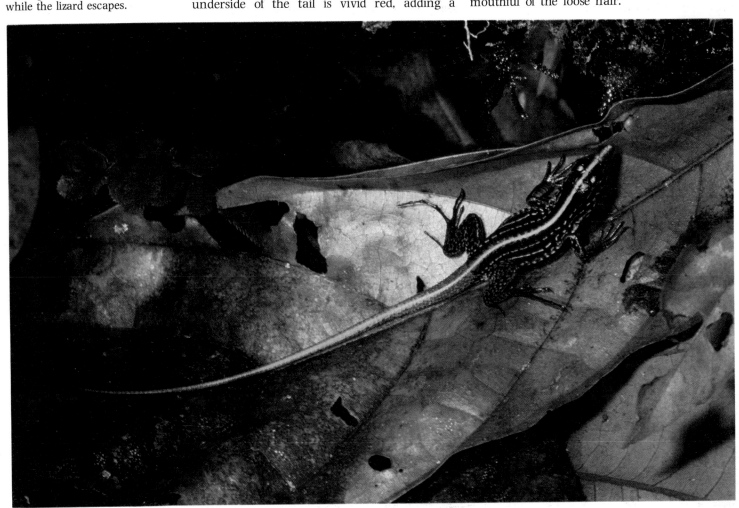

# Defence Routines

There can only be a few animals able to rely on a single defence strategy throughout their lives. The great majority make use of many different defences which they vary according to circumstances. One routine is that followed as an individual grows to maturity. Its problems early in life may differ greatly from those it will have as an adult, so as it matures it changes its defence techniques considerably. Then there are routines followed by adult animals, especially long-lived animals, with the passage of the seasons.

The simplest seasonal changes are those of camouflaged species that must alter their appearance to suit a varying background. This need is mainly restricted to Arctic birds and mammals. A more widespread reason for seasonal change is reproduction. Many birds, reptiles, amphibians and fishes temporarily divest themselves of concealing coloration in the breeding season. They adopt the bright nuptial dress they need to advertise themselves to rivals and opposite sex. Usually the male alone places his life in jeopardy this way, but in a few bird species, such as the painted snipe and various phalaropes, the roles are reversed; the female compromises her safety by being colourful while the male retains his camouflage. Where seasonal change in appearance is needed because of the annual moult, rather than for breeding, will be discussed later.

Nestling great tits 'freeze' if a predator looks into the gloomy nest cavity. The moss-green back and disruptive head-pattern is kept throughout life; adults need the same camouflage when incubating. Scrupulous nest-sanitation minimizes detection by smell.

The most common reason for a complete change of coloration in the life of an individual is an alteration in defence needs as it grows. The problems of a newly hatched youngster, or even more so of an egg, are quite different from those of an adult. For one thing it is much smaller, so a wider range of enemies are able to attack it. Also its weapons may not yet be developed, or if they are, they are not very effective. It must therefore rely on other forms of protection, particularly if it is not protected by one or both parents. Parental care of young is well developed only in birds, mammals and social insects (ants, bees, wasps and termites), though other animals may at least guard their eggs. Most young creatures are on their own and must have their own defences to meet their special needs. In many ways one can think of a newly hatched individual as a completely different 'species' from its parents. Apart from the size difference it often lives in a totally dissimilar environment and seeks food quite unlike that eaten by adults. Often its way of life is so totally different from that of mature individuals that in shape and pattern it bears little similarity to its parents. Indeed biologists often have a problem making the connection between a larval and adult form before the life-history has been worked out. Without prior knowledge who would guess that a legless white maggot, which needs no special colours since it lives inside its food, will become a shining bluebottle fly; or that a fluffy little camouflaged chick will one day become a gorgeously orna-mented cock pheasant? Obviously one cannot speak of the defensive adaptations of any species without specifying the stage it has reached. And as will be seen, there may be several different immature stages, each with a different set of solutions to its problems. In a butterfly life-cycle, for instance, the egg, caterpillar, chrysalis and adult are usually quite dissimilar in coloration. Moreover, even within the caterpillar stage, succeeding instars (the name given to an insect stage between moulting, of which there are normally five or more) may bear little resemblance to each other. Early instars of the pine-hawk moth caterpillar are green with longtitudinal stripes, making them cryptic among the pine needles. But the later instars, too fat to simulate the needles realistically, are brown and patterned like twigs. As their coloration changes, so does their behaviour. When small and green the caterpillars rest among the needles, whereas they lie along branches once they reach the brown stage.

Changes in appearance with age or season

are relatively slow, but at any time an animal may display a succession of defensive tactics which involve a changed appearance. This is because at any stage in its life it is unlikely to find one defence strategy effective for all forms of attack. This kind of quick routine is often referred to as a 'multiple defence system'. These repertoires have been evolved to cope with situations where the first line of defence fails. For instance, a well-camouflaged creature may be spotted, or stumbled on by a searching predator. It is then beneficial if it has a second defensive tactic, and a third if this also fails.

Simple multiple defence systems are found widely. The use of flash colours and intimidating false eyes already described fall into this category. They are normally employed only when camouflage – the primary defence – has failed.

It is in the tropics, however, and especially in rainforests, that the most complex examples of multiple defence systems are to be found among land animals. It is no coincidence that Henry Bates and Fritz Müller, the naturalists who did most to elucidate the principles of mimicry, both spent many long years in the Amazon forests. Also, the majority of modern specialists in protective coloration have all had extensive tropical experience. Certainly there are some beautiful examples of multiple defence systems in temperate forests, but the real eye-openers can be experienced only by those willing to face the discomfort of the jungle. It is interesting to speculate on the reasons for the complexity of tropical forest animals' defences. No positive answers can be given, but there is no harm in making a few informed guesses.

It is important to remember that tropical forests contain far more species than any temperate-zone habitats. They also harbour representatives of groups not found at all in cooler regions. One repercussion of this great diversity is that most animals have to defend themselves against an extremely wide range

Like most birds the lapwing changes its colour-pattern when it matures. As a chick (above left) it is protected by superb camouflage. The boldly marked adult (above right) leaves the nest at the first sign of danger and attempts to draw the intruder away with a 'broken-wing' distraction display.

of potential enemies. A camouflaged moth resting on tree bark anywhere is likely to be discovered by birds, but the variety of avian insectivores is greatest in the jungle, where it also risks attack by monkeys and lizards. Apart from the problem of direct attack, it can be flushed by a column of marauding army-ants (or driver-ants as they are called in Africa). Whether it flies or simply falls to the ground, it may be jumping out of the frying-pan into the fire. For many species of jungle birds methodically follow the army-ants as they move through the forest. Some stay above the ants to catch insects escaping by flying; others move across the forest floor ready to pounce on those that drop through the canopy. Because of such multifarious risks, many tropical animals have evolved multiple defence systems which aim to cope with as many as possible. Needless to say, the complex adaptations of the more vulnerable prey species have been matched by the predators becoming increasingly specialized for finding and dealing with them.

This leads us to the other reason why the defensive adaptations of tropical species are often so extreme. In the past, Northern Eurasia

Right: The arctic hare, like many other arctic mammals, undergoes a seasonal colour change. This maintains the effectiveness of its camouflage throughout the year.

and North America have been greatly affected by widespread glaciation, and many species that were evolving complicated defensive repertoires were undoubtedly wiped out. Close to the equator, conditions have been relatively stable for millions of years so that all manner of extreme specializations have had time to evolve.

Having established the general features of defence routines, we can take a detailed look at some examples, beginning with seasonal changes in coloration by adult animals. The annual change in the colour of many birds and mammals has already been reviewed in the section on camouflage, but additional comment can be made concerning the timing of the change. Neither birds nor mammals can alter their external appearance rapidly since feathers and fur are dead structures. Colour-change necessitates a moult which involves shedding the old coat and the growth of a new one. Moult is a demanding process requiring the synthesis of large quantities of keratin, the special protein of which feathers and hair are made. It also involves changes in the skin, and even the blood supply has to be augmented to meet the increased growth demands. Furthermore, since the covering is needed for thermal insulation, it cannot be cast off at once.

Moult is therefore a slow process taking many weeks to complete. This means that ptarmigan, hares and Arctic foxes must begin the change to a white coat in autumn before there is a snow cover. They may therefore be relatively conspicuous for a while, though this is probably a small price to pay compared with the advantages gained in winter by the colour adjustment. In spring the problem is less acute since patches of snow remain unmelted till quite late, so brown and white animals in mid-moult do not stand out.

The most widespread cause of seasonal colour-changes is not the need to suit a particular background, but to satisfy the special demands of reproduction. Most animals that make contact with one another visually use colour in their mating displays. If they are social they may also use colour as a means of expressing their position in a hierarchy. For the more vulnerable species with many enemies the competitive value of showy colours poses a dilemma. They need to be as inconspicuous as possible if they are to escape the attention of enemies, yet in their social life they can score over rivals only by being colourful. As we have seen, one way of combining the two conflicting demands is to be generally cryptic, and have bright colours on a part of the body

coloured feathers on the head and body for this difficult time. This 'eclipse' plumage affords them some camouflage while they are hiding in the reeds waiting to be able to fly again. As soon as the new flight feathers are functional, the eclipse body plumage is replaced by the definitive male dress. The ducks are cryptically plumaged throughout the year and therefore do not require a special eclipse plumage.

The degree to which young animals differ in coloration from adults varies greatly, even within a particular group. Among mammals the youngsters are usually small editions of their parents by the time they are fully active. Baby mice, squirrels and other small mammals born in safe nests may have a period of near-nakedness at the beginning of their lives, but this cannot be regarded as protective coloration. It is simply that they are born in a very immature state of development when they are still hairless. By the time they emerge from the nest they have a full covering of hair and usually look like their elders. Most immature mammals have been able to dispense with a specially protective juvenile coat because they are looked after by the adults until they are well grown. There are a few exceptions, such as deer fawns with their spotted coats, and the cheetah kitten with its possible mimicry of the ratel, commented on earlier. Also young pigs and tapirs have camouflaged spotted coats as protection until they gain their adult strength and weapons.

Left: Only for the brief nesting season is the male little bishop brightly-coloured. For the rest of the year he is an inconspicuous dull brown bird like the female.

Below: Unusual among mammals, wild pigs have a distinctive juvenile pattern which is cryptic. They lose it when they grow strong enough to defend themselves actively.

where they can be exhibited only when necessary. The other solution, which falls under the heading of a routine, is to wear the bright colours for only a brief period of the year when pair formation is taking place. Fishes, amphibians and reptiles can assume sexual colours rapidly by the relatively simple process of colour adjustment in the skin. For birds, however, the change is slow and generally requires a moult of the body feathers. This 'pre-nuptial' moult is additional to the complete moult after the breeding season. The males of Arctic-breeding waders such as turnstones, godwits, and many sandpipers adopt a bright breeding plumage which is lost soon after the brief nesting period. Similarly, in tropical Africa the males of many weaver-birds change into a flamboyant nuptial dress at the start of the rains; the rest of the year they are drab like the females. The little bishop, *Euplectes orix*, which is a good example, gets its name from the vivid scarlet-and-black plumage of the breeding male.

Unlike most other birds, many of the ducks and geese perform such a rapid moult of their wing feathers that they become temporarily flightless — and therefore vulnerable to predators like falcons. The drakes of some wildfowl, such as mallard, teal and eider, which are brightly plumaged, develop a covering of drab-

With birds a distinctive juvenile pattern is the rule rather than the exception. The European robin, for instance, is brown all over and spotted like a thrush when it begins independent life. Its camouflage helps it survive the rigours of early independent existence. All being well, by the time the red breast feathers appear and make it more conspicuous it has sufficient experience of cats and other enemies to avoid them. Before the dull juvenile plumage, which is fully functional for flight, there is often another of soft down feathers worn while the baby bird is still in its parents' care. This is the nestling plumage and serves to hide the youngster in the weeks before it can fly. A cryptic nesting plumage is particularly important for so-called 'nidifugous' young like those of pheasants, ducks and gulls that leave the nest soon after hatching. Those that remain in the nest, called 'nidicolous' young, may be almost naked or have thin wisps of nestling down, for they are protected inside the nest which is well concealed or placed out of predators' reach. The cryptic attributes of some birds' plumage is difficult to appreciate when we see them going about their daily lives. The great tit, for example,

does not look particularly cryptic, yet when sitting in its mossy nest at the bottom of a hole in a tree the tit's green back merges well with the background, and its black-and-white striped head is extremely disruptive. The nestling tits share the same cryptic patterning and help the crypsis by keeping the nest clean. They defecate only when the parent is present, so that it can carry the neatly-packaged pellet away from the nest-hole. For these titmice the main danger is of a weasel looking into the hole, seeing or smelling birds inside, and climbing in to eat them. This is when the camouflage may save their lives.

Among reptiles colour-changes associated with age are not profound as a rule. Most species are cryptic at all stages though some males develop bright sexual colours as they mature. Curiously there are a few species in which the usual trend is reversed, young individuals being much more colourful than the adults. Some Old-World skinks, and tropical American lizards, like *Ameiva*, have brilliantly coloured tails, often bright blue, which serve to deflect attacks towards that end of the body. Some species retain this feature when they

Below: The cinnabar moth is protected throughout life by poisons it derives from ragwort, the larval food-plant.
The warning colours of the caterpillars (left) are completely different from those used by the adults (right).

grow up, but others lose it and become drab all over. In such cases a possible reason for the change is that the youngster needs to move about a great deal in search of small prey. As their frequent movement spoils the camouflage they have a deflective tail as a further safeguard. When fully grown, however, the need to feed actively may be less so that camouflage alone best serves their defence needs.

Frogs, toads and other amphibians undergo a profound change of appearance when they metamorphose from the fish-like tadpole stage to a terrestrial quadruped. As adults they are either highly cryptic or vividly coloured, depending on their palatability. The young tadpoles, however, are mostly black and highly conspicuous. This applies even to tadpoles of palatable species such as the common European frog, *Rana temporaria*, which seems rather odd since they possess no obvious noxious attributes. They are quite conspicuous and would seem to be easy prey for predatory fish and birds. Possibly the newly hatched frog tadpoles mimic the tadpoles of *Bufo* toads which are protected from enemies by poisonous skin secretions. An alternative, physiological ex-

planation for the dark colour of tadpoles is that it serves mainly to absorb radiant heat from the sun.

Among land-dwelling animals insects exhibit the most spectacular change in form and colour associated with development. Larval and adult insects often lead such completely different lives that any great similarity in their appearance would be surprising. The larva is adapted to find and consume as much body-building nutrients as possible, while the adult's main task is to locate a partner and produce fertile eggs. Because their ways of life are so dissimilar they are liable to meet different enemies, and therefore need different defensive adaptations.

Some well camouflaged insects are equally cryptic as adults or larvae, but since their backgrounds are not the same they wear different colours. Many of the brown moths that rest by day concealed on tree bark are bright-green as caterpillars, this being appropriate to their background of fresh leaves.

Adult insects that display warning colours, indicating they are unpalatable, are often coloured strongly when they are larvae,

Left: The European alder moth *Apatele alni* provides a good example of changing defence strategy with age. In its early stages, the caterpillar is disguised as a bird dropping (above), but the last larval stage (below) has a conspicuous (probably bluff) warning pattern. This provides protection when it is actively searching for a place to pupate. The adult moth (right) rests on tree bark and is protected by camouflage.

though the colour combination may differ. The cinnabar moth, *Callimorpha jacobaeae*, a common European species, has red-and-black wings and a red abdomen, making it conspicuous as it flies about during the day. The caterpillars have no red, but are equally colourful by virtue of transverse black-and-yellow bands the whole length of the body. Similarly, the American milkweed butterfly is as conspicuous at the caterpillar stage as it is when it becomes a butterfly, though the warning is given by different colours. In both these species the insects contain at all stages poisons derived from the larval food plant, stored and used to protect subsequent stages.

The last examples are uncomplicated in that both adult and larva rely on the same primary defence strategy. But often the story is vastly more complicated. And one does not have to go to the tropics to see fascinating examples of changing defence strategies in the life of an insect. In European oak woods the alder moth, *Apateles alni*, changes from disguise to mimicry during its caterpillar phase. In its early instars it is a perfect copy of a bird-dropping; it has a black-and-white pattern and sits curled in the middle of a leaf just like the real thing, moving infrequently so as not to give itself away. But the last instar caterpillar appearing after the last skin-change looks completely different, being marked with alternating bands of black and chrome-yellow. Such profound transformations are not rare; many hawk-moth and swallowtail butterfly larvae are also disguised as bird excrement when small, then change to another disguise or camouflage as they become too large to simulate a dropping convincingly. What is particularly interesting about the alder is the reason for the switch from disguise to advertisement. As a growing caterpillar it need never move far since it is surrounded by food, but the last larval instar must eventually leave the foliage completely in order to pupate. It moves along the branches seeking a rotten branch in which to burrow, then it must spend several more hours excavating a tunnel into the wood, and a chamber at the end of this. To do all this it must obviously move about a great deal and if it retained its bird-dropping disguise till this stage it could easily fall prey to a hungry bird not fooled by a dropping that walks. Instead, this last stage bears typical warning

colours, and when only its hind end is seen as it bores into the wood it is quite wasp-like.

Another intriguing routine is followed by a stick-insect from New Guinea, *Extatosoma tiararum*. At the egg stage it looks like a dry seed lying on the forest floor. Out of the egg comes a dark, fast-moving nymph, looking and behaving like a vicious tailor-ant; this is its second disguise. Later, when it grows too large to resemble an ant, it changes to a pale-grey colour. Instead of rushing about like an ant it now moves deliberately, the long tail arched over its back like a scorpion – disguise number three. Its fourth and last imitation, adopted when adult (and by now a large insect with wings), is as a dried-up fragment of vegetation, a more normal disguise for a stick-insect.

'Multiple defence systems' refers to routines of a quite different kind in which a succession of defensive tactics are tried by an animal under attack. Most animals have several defensive techniques at their disposal at any stage in

their lives, each appropriate to a particular enemy or circumstance. Some are directed towards predators which hunt by smell or touch and involve no obvious use of colour. But even if defences aimed at foiling predators that hunt by sight are alone considered, one may still find that a particular animal has several cards it can play. The following examples show how multiple defence routines operate.

The large brown caterpillars of an East African moth belonging to the Lymantriidae family feed at night on the foliage of tall trees. Before dawn each day they descend the trunk and assemble near ground level as a tightly packed mass on the rough bark. The caterpillars' overall coloration is mottled brown, like the bark, and each caterpillar has hairs on its flanks which overlap neighbours lying alongside. The overall effect is an irregular patch indistinguishable at any distance from the surface of the tree trunk. Left alone they remain immobile throughout the day, depending on their camouflage to escape detection. But if disturbed, say by a bird methodically searching the bark for insects, they immediately react by adopting their second defence position. Each caterpillar fluffs out its hairs, making itself look

larger, arches its thorax to reveal prominent tufts of red spines previously hidden and secretes a large drop of green fluid which remains hanging from its mouth. This warning display is probably sufficient to deter many would-be enemies, but the caterpillars can do more if the marauder begins to attack them. First, they begin to wave their front ends about wildly. This probably serves to bring the tufts of loose red hairs, which are barbed and probably coated with an irritant substance, into the attacker's face. Then, as a further bid to save themselves from a really persistent enemy such as a cuckoo, the caterpillars drop one by one to the ground. There, as a fifth and final manoeuvre, they wriggle furiously so as to cover themselves with dead leaves. By the time the attacker has consumed a few of the larvae slow to drop, and flies down to find the rest on the ground, many at least are hidden from view. In this example the sequence is first camouflage, followed by warning, physical attack and finally attempted escape. A similar sequence is followed by the European puss moth caterpillar. Normally it is well camouflaged as it feeds upside-down among willow leaves. It has a disruptive pattern of green and dark-brown and inverted countershading. If molested,

When it is alarmed the non-venomous hog-nosed snake first mimics a rattlesnake (below). If the threat persists, it then turns on its back and feigns death (right). The 'bad meat' colour of the mouth and cloaca, with an accompanying foul stench, make the corpse-display most convincing.

it turns a brilliant red 'face' towards the intruder and extrudes two long red filaments from its tail. At the same time an odour of formic acid can be detected. As a last resort formic acid is actually sprayed into the face of an attacker.

The American hog-nosed snake, *Heterodon nasicus*, also begins its routine with camouflage, but its second line of defence is mimicry. If despite its background coloration it is disturbed, this harmless snake immediately puffs out its head to resemble one of the much-feared venomous rattlesnakes found in the same habitat. It even imitates the rattlesnake's intimidating rattle by rasping the scales along the sides of its body. If this mimicry fails to send an enemy scurrying away, the hog-nosed snake moves to its third line of defence — feigning death. It turns over on to its back and lies quite still with its mouth wide open. The naked-looking cloacal region is everted, as if pushed out by gases of decomposition inside the body. The lining of the mouth and the cloaca are coloured purplish-red, the colour of rotting meat, and a foetid odour is emitted to make the corpse effect even more convincing. All being well, even a predator that took no notice of the rattlesnake display stops short at eating something that looks and smells as if long dead. It might seem unlikely that a bird or mammal predator would be fooled this way, for the snake was obviously alive a few moments before. But few animals are capable of rational thought; they simply react instinctively to the set of stimuli presented to their senses at any one time. With the exception of a few specialized scavengers, most carnivores are averse to touching flesh which is rendered unpalatable — even poisonous — by putrifying bacteria.

# Aggressive Use of Colour

In view of the widespread use of concealing coloration for defence, it is hardly surprising that some predators have also adopted it – for offensive purposes. Hunters by stealth may rely greatly on camouflage in order to approach their victims unseen. For the nearer a killer can get to its victim before being spotted, the shorter will be the final dash, and greater the chance of securing a meal. Even more reliant on concealment are those less active predators that wait in ambush for their prey. Many have evolved convincing disguises as inanimate objects, or more sinister yet, as one of their victims' food items. Mimicry also is employed by some predators as a means of approaching their target unnoticed. An example will be given later where one small species of fish imitates another in order to take a bite out of the large fish with which the model amicably consorts.

It would be wrong, however, to assume that all predators with concealing coloration have evolved it as an aid to hunting. In the case of powerful animals such as lions and tigers the situation is unambiguous. But with smaller predators which can themselves be predated, crypsis may be dual-purpose. Indeed, for some predators such as chameleons, camouflage probably contributes little to hunting success, but is of great protective value.

Hunters-by-stealth are generally well camouflaged or disguised and have the ability to move

Small predators, like this tree-snake *Dendroaspis*, may benefit from concealing coloration when they are lying in wait for their prey. However, the protection it affords from their own enemies may be of greater importance for their survival.

towards their prey without their body movements being perceived. Perhaps most familiar are members of the cat family, including the domestic cat. Varieties of all shades and patterns have been selected by man showing how much genetic latitude there can be in a species, but the wild cats from which they are derived are beautifully camouflaged. What we call the 'tabby' pattern in domestic cats comes closest to the original. In the Felidae (cat family) generally there are two main categories of coat colour: there are plain, sandy-coloured species, of which the African and Asiatic lions are best known; other representatives are the caracal, or desert lynx, of Africa and the American mountain lion, or puma. These are dry-country animals which blend well with their background for most of the year. During wet periods their camouflage may become ineffective, but they can then often conceal themselves in long grass. In the second category are the more numerous strongly patterned species of cat. Some are striped, like the tiger, American bobcat and European wild cat, while others are spotted. Essentially these are just two forms of disruptive patterning which serve to break the hunter's outline. The spotted cats include the cheetah and the serval, which inhabit open country and have small spots, while forest-dwellers such as ocelot and jaguar are heavily spotted. Each species is adapted to different kinds of dappled sunlight in the places where it rests or hunts.

The leopard is a particularly interesting species in this context. In Africa and the Middle East, where most leopards live in open country, the majority are pale-yellow animals with black spots. All-black (melanic) individuals do occur from time to time, by chance mutation, but they are rare. Supposedly, in

this kind of country black leopards are at a strong disadvantage when hunting gazelles and other quick-sighted prey. They are therefore selected out of the population. In the jungles of South-east Asia, however, the melanic form prevails and 'typical' leopards are rare. The black leopards are the 'black panthers' that many erroneously believe to be a distinct species. Natural selection has favoured black individuals in rainforest situations, presumably because they are the most effective hunters in dark habitats. Curiously, the tiger has not diversified in a similar fashion. Tigers living in the permanent gloom of Malayan forests have the same striped pattern as tigers in India and China. Their disruptive pattern may seem obviously to be adapted to a background of elephant-grass, but there is virtually no ground vegetation at all in the jungle. It would appear, then, that the tiger's stripes constitute not so much background picturing as disruptive patterning, suited to a range of dissimilar habitats.

Left: The complicated pattern of the leopard (above) and the plain coat of its arctic relative the Canadian lynx (below) both provide concealment when hunting in their vastly different habitats.

Below: The Nile crocodile's resemblance to a floating log enables it to approach animals standing in shallow water without being detected.

Overleaf: African lions rely greatly on their sandy coloration when hunting in the dry season. The background becomes green during the wet season but tall grasses then provide adequate cover.

Not all carnivorous mammals are camouflaged like the cats. The African hunting-dog, for example, has a highly conspicuous blotched pattern which presumably has a special meaning in the social context. Hunting-dogs are highly gregarious animals with an intensely social life. Camouflage would be of little use to wild dogs, anyway, for they hunt as a pack. Instead of approaching the victim stealthily they give chase openly and pursue it relentlessly until it is exhausted. Similarly, wolves are pack animals relying little on concealment when hunting.

Barracuda and many other predatory fishes in the sea exhibit typical concealing coloration which must be of value for hunting, but since they are liable to be predated on themselves by larger fishes it is difficult to be sure of its primary function. However, the blue-grey tones and countershading of many ferocious sharks doubtlessly serves primarily for aggression purposes.

In rivers, predators can make use of the current to bring them towards an unsuspecting prey. They can thus make a stealthy approach while remaining motionless, or nearly so. Such predators make use of disguise as inanimate objects, a form of concealment well suited to this form of attack. What, for instance, could be more normal than a dead leaf floating downstream? Small fishes must find it so unexceptional that they do not take fright even if the leaf passes them very closely. In behavioural parlance, they become 'habituated' to this frequent, harmless experience. The South American leaf-fish, *Monocirrhus polyacanthus* (not to be confused with the marine leaf-fish *Platax*), has taken advantage of this situation by evolving an extremely convincing disguise. Though only three inches long, it is a voracious predator able to swallow fishes almost as big as itself by a sudden opening of its cavernous mouth. In shape and colour it is a perfect copy of a dead leaf, complete with simulated mouldy patches, a mid-rib and a serrated edge. It even has a projection from its chin that looks like the leaf's petiole. When hunting, the leaf-fish hangs head down below the water surface, letting

itself drift slowly towards its intended prey. Alternatively, in still water it can propel itself slowly toward its victims by imperceptible movements of transparent pectoral and tail fins.

Another object frequently to be seen floating down a river is a length of broken branch or log. Again we find that a predator, in this case a reptile, has developed an appropriate disguise. Large crocodiles and alligators can bear an uncanny resemblance to a floating log, while young specimens are like small pieces of wood. This disguise permits these powerful creatures to approach animals drinking at the water's edge, or standing in the shallows, and with a final, swift lunge the jaws are clamped on the luckless animal's head or a limb. It is then dragged into deeper water where it soon drowns and can be dismembered.

Moving now to predators that lie in wait for their victims, we can draw on a wide choice of examples. Though many snakes are active hunters quite a number are sluggish animals that wait in ambush for their prey. Those pythons, vipers and boas that wait on tree branches, or hidden among the ground-litter, generally have a variegated pattern of creams and browns that is exceedingly disruptive. Despite the great size some of them reach, they are almost impossible to make out when they are still. They can remain immobile for hours on end – even for days sometimes. But if a potential prey comes near enough the initial strike is of lightning speed. If a viper or other venomous species is involved this is all that is required. The victim quickly succumbs on the spot, or moves lamely a short distance and is followed by its attacker. If it is a non-poisonous boa, or python, the initial strike is swiftly followed by coiling of the body around the victim in a lethal embrace. In the canopies of evergreen tropical forests live many slender green snakes belonging to various families. The green mamba is the most notorious in Africa, because of its venomous bite, while the green pit-vipers in the genus *Bothrops* are equally feared in the American tropics.

The efficacy of snakes' camouflage cannot be in doubt and it must surely aid their capture

Below: The angler fish captures prey attracted to the fleshy lure held by a filament over the cavernous mouth. Victims fail to recognize the predator because of its disguised body-form.

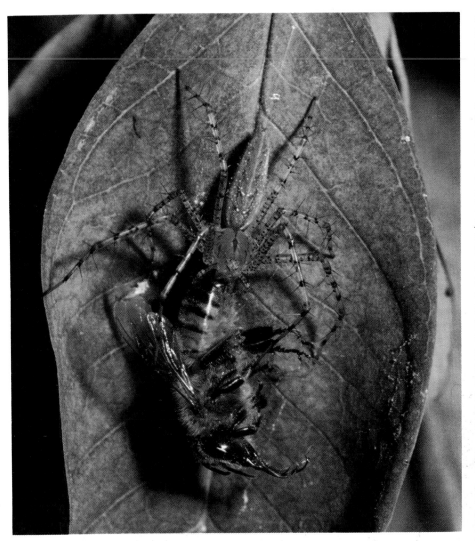

South American lynx spider with captured bee. The spider's green colour helps it seize prey and camouflage itself from its enemies.

rocks over with the hands. The sting is so excruciatingly painful that when a fisherman is stung while sorting his catch it is recommended that he be tied down, lest he throws himself overboard in frenzy.

Ambush is a widespread hunting technique among lynx- and crab-spiders, praying mantises and other invertebrate predators. Some of them exhibit remarkable camouflage or disguise, but this may be needed less for feeding than for avoiding being fed upon. There are, however, a few species which have certainly adapted colour to their particular way of catching prey. The bird-dropping spider, *Ornithoscatoides decipiens*, of South-east Asia, spins a film of white silk in a small patch on the upperside of a leaf. It then sits at the centre of the pad, its body disguised like the solid portion of a bird-dropping, the shiny silk area representing the usual wet accompaniment. Flies and other insects attracted by fresh droppings, on which they feed, come to the spider which promptly seizes them. Some butterflies and moths have a predillection for faecal matter and urine as a source of nitrogen; the bird-dropping spider catches these also. Since the discovery of this spider in Java last century, other species showing the same adaptation have been discovered elsewhere in the tropics.

Another instance of 'alluring coloration', as it is sometimes called, is provided by the flower-mantis, *Hymenopus bicornis*, from the Malaysian Archipelago. This beautiful insect is shaped and tinted like a delicate flower, some being white and others a deep pink. Even the mantis's legs have expanded sections resembling petals. If the flower mantis simply waited on green foliage, attracting flies to itself like the bird-dropping spider, the value of its disguise would be clear. But it does not. Like the African flower mantis, *Pseudocreobotra*, described in an earlier section, *Hymenopus* takes up its waiting pose inside an open flower. Therefore its allurement may well be combined with a measure of defence against insectivorous birds.

The spider described above could be said to use its own body as a lure to bring its victims within reach. Other predators are known to use a special organ as a lure; often it resembles their victims' preferred food items. Angler-fishes are ugly, bottom-living fishes which derive their name from a modified fin-ray projecting above the head like an angler's rod. At its tip is a fleshy mass which serves as the bait and when another fish comes to investigate the bait, which hangs above the huge mouth, it is seized and swallowed. Some angler-fishes have virtually abandoned swimming and move about the sea floor on reduced, limblike fins. Deep-sea angler-fishes living in total darkness have luminescent lures which can be made to wriggle like worms.

of birds, monkeys and other prey with acute vision and quick reactions. Nevertheless, it is also true that many snakes require concealment for defence, for in environments where snakes abound there is usually at least one snake-eating specialist, and often there are many.

The scorpion-fishes are so called because of a stout and often poisonous spine in the dorsal fin. They occur in shallow, warm seas and are all carnivores; for the most part they are sedentary fishes camouflaged like the rocks and some are convincingly disguised as lumps of old coral with filamentous outgrowths of the skin looking like encrusting seaweeds. The stonefish, *Synanceja horrida*, is a member of this family and well known to divers. Its method of feeding is to lie motionless on the sea-bottom with the appearance of an eroded piece of coral; when an inquisitive fish comes near it is swallowed with surprising speed. The dorsal spine is defensive; it normally lies flat but can be erected rapidly when required. Stonefishes are often plentiful and one of the reasons why it's unwise to walk barefoot on a reef, or turn

113

Right: A flower mantis from Borneo. The bright colour and petal-like expansions of the body and legs attract flower-visiting insects which are seized and eaten.

Another animal with a worm-like lure is the alligator snapping turtle, *Macrochelys temmincki*, of the lower Mississipi and other rivers in south-eastern U.S.A. A large animal, reaching 100 kg in weight, it can nonetheless make itself quite inconspicuous, for its rough shell carries a dense growth of green algae. The 'alligator snapper' lies on the river-bottom with its mouth wide open. The inside is darkly coloured, but on the floor of the mouth is a fleshy filament, bright-pink in colour. This can be made to wriggle by muscular action. Fishes have been observed to swim into the turtle's mouth attracted by the lure. Though generally sluggish these huge reptiles well deserve their colloquial name for they can shut their jaws with surprising speed and great force. On land, snapping turtles can be quite aggressive, unlike most of their relatives.

Just as deep-sea angler-fishes have luminescent lures for attracting prey in the dark abyssal depths, there is a predatory insect using a similar ploy to attract night-flying prey. The larva of a fungus-gnat, *Arachnocampa luminosa*, from New Zealand, inhabits cave entrances where it feeds on insects, unlike typical members

of the family which develop in mushrooms. *Arachnocampa* larvae spin traps of fine threads, some of which hang vertically downwards and bear sticky droplets. At night each larva illuminates its trap with 'living light' produced chemically by a luminous organ. Small insects are attracted, become stuck on the sticky strands and are eaten by the larva.

The story of the cleaner mimic is surely one of the most fascinating in a subject not short of surprises. The models for the mimicry are small species of wrasse (a mainly carnivorous group of fishes) living on coral reefs. Like certain shrimps they have become specialized cleaners, removing ectoparasites from fishes of all types and sizes. Cleaner wrasse are only a few inches long, yet they can swim close to, and even into, the mouths of normally pugnacious predators. The fishes being cleaned benefit by ridding themselves of ectoparasites with which they are constantly plagued; in turn the wrasse are sure of a constant and plentiful food supply. During the evolution of this symbiotic relationship the cleaners have developed a characteristic appearance and behaviour. Their pattern of black stripes on a blue or yellow background is recognized by hundreds of reef-inhabiting fishes. The cleaners further identify themselves by a characteristic dancing movement as they swim up to each 'customer'. Individual cleaners, or pairs, have definite territories where they are visited by large numbers of fish every day. Almost unbelievably, queues of fish have been seen waiting their turn at a cleaning station!

One of these cleaner wrasse, the 'sea swallow' or 'blue streak', *Labroides dimidiatus*, is mimicked by the sabre-toothed blenny, *Aspidontus tractus*. The blenny has evolved almost precisely the same slender shape and colour-pattern as the wrasse, though it isn't closely related. Moreover, geographical races of the cleaner wrasse, which differ slightly in pattern, have each been copied by a race of the blenny. The special feature of the mimic, as its name tells us, is its sharp teeth with which it takes bites out of other fish. The sabre-toothed blenny does not kill its victims, which are often huge by comparison; it simply removes pieces from their fins. The blenny approaches a large fish with the same dancing movement used by the cleaner wrasse. This and the familiar pattern fools many of the fish attending the cleaning station so that they remain still while the blenny dances up to them. Then it takes its bite and makes off at speed.

One might expect that fish living on the reef would learn to distinguish between the two species, one of which gives comfort and the other pain; and it seems that older inhabitants do. They have been seen to make deliberate attempts to avoid the attentions of the blenny while retaining the cleaner's useful services. It is the young, inexperienced individuals that bear the brunt of the blenny's insidious attacks.

Finally there is the predator that mimics its own prey, not in order to capture them (which is easy), but to foil the victims' protectors. The predator is the nymph of a lacewing, *Chrysopa slossonae*, which lives in colonies of the woolly alder aphid, *Procipilus tesselatus*. The nymph feeds on the aphids, picking them up one by one and sucking them dry. Before discarding the shrivelled remains it picks the waxy tufts from the aphids and fastens them to its own back. So thoroughly does it encrust itself with the wax that it becomes indistinguishable from them. Were this all, one might suggest that the lacewing nymph required the mimicry as protection from birds which seem not to favour wax-covered bugs, but it has been proved by experiment to have a different role. The aphids are attended constantly by ants that seek their honeydew secretion. As long as a lacewing nymph is clothed in the aphids' wax it is ignored, but if cleaned of the coating it is immediately thrown off the plant by the ants.

The last account does not come from the diary of a nineteenth-century naturalist. It is a recent discovery, made in New York State, and documented in 1978. This underlines the fact that there remain to be discovered countless thousands of unusual ways in which colour, pattern and form are combined by animals for their survival. Virtually no special equipment is needed by the investigator — just interest and plenty of patience.

Right: The cleaner wrasse provides a valuable service to large reef fishes by removing small parasites from their skin and gills.

Figure above: The characteristic form and pattern of the cleaner (below) is closely mimicked by the sabre-toothed blenny (above) which takes bites from the skin of the fishes expecting to be cleaned.

# Bibliography

Brower, L. P. Ecological Chemistry (Scientific American 220 pp 23–29 1969)

Cott, H. B. Adaptive Coloration in Animals (Methuen, London 1940)

Edmunds, M. Defence in Animals (Longman, London 1974)

Fogden, M. & Fogden, P. Animals and their Colours (Peter Lowe, London 1974)

Fox, H. M. & Vevers, G. The Nature of Animal Colours (Sidgwick & Jackson, London 1960)

Kettlewell, H. B. D. Brazilian Insect Adaptations (Endeavour 18 pp 200–210 1959)

Owen, D. F. Tropical Butterflies (Clarendon Press, Oxford 1971)

Wickler, W. Mimicry in Plants and Animals (Weidenfeld & Nicholson, London 1968)

# Acknowledgements

The photographs in this book were taken by the author, with the exception of those on pages:

18 M. P. L. Fogden; 21 (bottom) S. C. Bisserot/Bruce Coleman; 27 (top) M. P. L. Fogden; 31 Allan Power/Bruce Coleman; 32 Jane Burton/Bruce Coleman; 33 M. P. L. Fogden; 43 M. P. Harris/Bruce Coleman; 46 (top) Jon Kenfield/Bruce Coleman (bottom left) Jon Kenfield/Bruce Coleman; 47 P. Laboute/Jacana; 49 Leonard Lee Rue III/Bruce Coleman; 53 (bottom left) M. P. L. Fogden; 55 (top right) J. L. S. Dubois/Jacana; 59 Leonard Lee Rue III/Bruce Coleman; 64 D. Faulkner; 66 I. Wyllie, Monkswood; 69 (bottom) Fievet/Jacana; 70 (top) Lee Lyon/Bruce Coleman (bottom) Norman Myers/Bruce Coleman; 71 I. Wyllie, Monkswood; 72 (top) M. P. L. Fogden; 77 Natural Science Photos; 82 M. P. L. Fogden; 83 (bottom) S. and K. Breeden; 87 M. P. L. Fogden; 92 Jane Burton/Bruce Coleman; 97 Gordon Langsbury/Bruce Coleman; 98 Jen and Des Bartlett/Bruce Coleman; 99 (bottom) Arthus-Bertrand/Jacana; 104 and 105 M. P. L. Fogden; 106 R. König; 108 (top) Goetz D. Plage/Bruce Coleman (bottom) Charlie Ott/Bruce Coleman; 112 Herré Chaumeton/Jacana; 114 M. P. L. Fogden; 115 (bottom) Bill Wood/Bruce Coleman.

# Index